REVISE AQA GCSE (9–1)
Biology

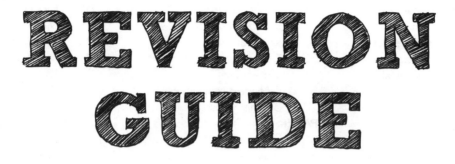

REVISION GUIDE

Higher

Series Editor: Harry Smith

Author: Nigel Saunders

I don't know who's gonna get this but u will get 9!

Also available to support your revision:

Contents

A small bit of small print:
AQA publishes Sample Assessment Material and the Specification on its website. This is the official content and this book should be used in conjunction with it. The questions in Now try this have been written to help you practise every topic in the book. Remember: the real exam questions may not look like this.

Microscopes and magnification

Elements

The **light microscope** was invented about 350 years ago. It uses light to form magnified images of cells and other tiny objects.

The best light microscope can produce magnifications of about 2000 times: ×2000.

The **electron microscope** was invented in the last century. It uses electrons to produce magnifications of up to 2 million times.

The higher magnification and resolving power of electron microscopes allow scientists to see and understand sub-cellular structures.

Resolving power

A microscope with a high **resolution** allows you to distinguish between two objects that are very close together.

Microscope	Resolution (nm)
light microscope	200
electron microscope (scanning)	10
electron microscope (transmission)	0.1

🖩 Maths skills   Magnification

You can calculate **magnification** using:

$$\text{magnification} = \frac{\text{size of image}}{\text{size of real image}}$$

You can rearrange the equation to find the size of the real object, if you know the magnification and the size of the image:

$$\text{size of real object} = \frac{\text{size of image}}{\text{magnification}}$$

Worked example

A light microscope has a ×10 eyepiece lens and a ×40 objective lens. The image of an onion cell viewed using this microscope is 0.2 mm long.

(a) Calculate the total magnification of the microscope.   **(1 mark)**

total magnification = 10 × 40 = ×400

(b) Use your answer to part (a) to calculate the real length of the cell.   **(1 mark)**

$$\frac{\text{size of image}}{\text{magnification}} = \frac{0.2\,\text{mm}}{400} = 0.0005\,\text{mm}$$

1 mm = 1000 μm
0.0005 × 1000 = 0.5 μm

Scale bars

Diagrams and photos of microscope images often have a scale bar on them.

A scale bar lets you:
- easily see the real sizes of the different objects in the diagram or photo
- calculate the magnification, if you do not know what this is.

If you want to calculate the magnification, you need to use a ruler to measure the real length of the scale bar.

It is better to make your measurements in millimetres, mm (rather than in centimetres, cm).

Worked example

The diagram shows a cell with a scale bar.

← 2 cm

1 μm

Calculate the magnification of this image. **(2 marks)**

length of scale bar on diagram = 20 (mm)

= 20 × 1000 = 20 000 (μm)

real length (from scale bar) = 1 μm

$$\text{magnification} = \frac{20\,000}{1} = ×20\,000$$

Now try this

A light microscope has a ×5 eyepiece lens and a ×20 objective lens. The image of a cell is 15 mm long. Calculate the actual length of the cell. Give your final answer in μm.   **(3 marks)**

$$\text{real} = \frac{15 \times 1000}{100} = 150\,\mu m \checkmark$$

1

Animal and plant cells

Animal **cells** and plant cells contain smaller structures which have different functions.

Structures of animal cells and plant cells

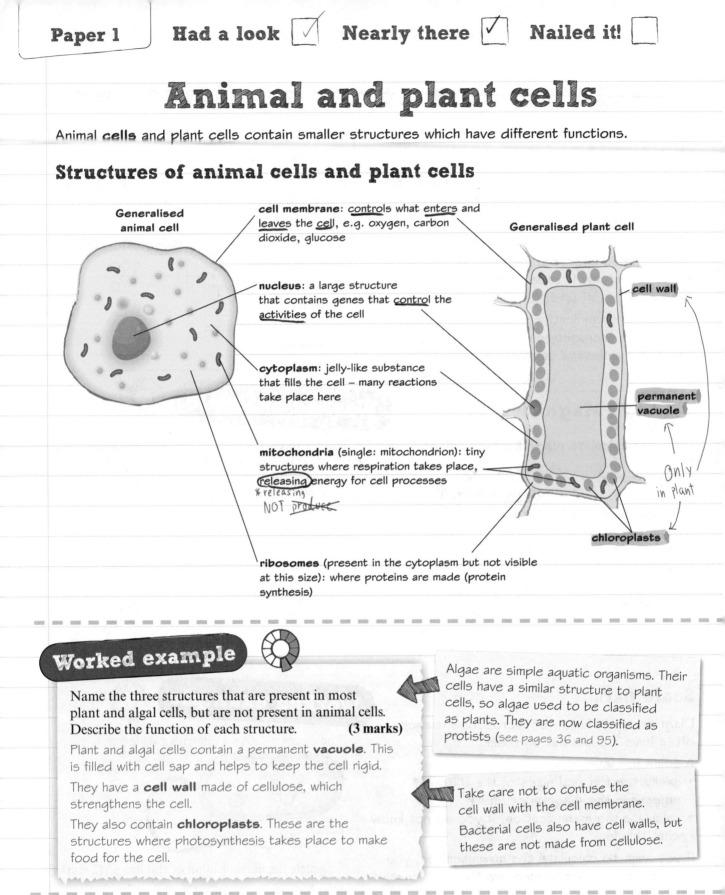

Generalised animal cell

cell membrane: <u>controls</u> what <u>enters</u> and leaves the <u>cell</u>, e.g. oxygen, carbon dioxide, glucose

nucleus: a large structure that contains genes that <u>control</u> the <u>activities</u> of the cell

cytoplasm: jelly-like substance that fills the cell – many reactions take place here

mitochondria (single: mitochondrion): tiny structures where respiration takes place, releasing energy for cell processes

* releasing
NOT produce

ribosomes (present in the cytoplasm but not visible at this size): where proteins are made (protein synthesis)

Generalised plant cell

cell wall

permanent vacuole

Only in plant

chloroplasts

Worked example

Name the three structures that are present in most plant and algal cells, but are not present in animal cells. Describe the function of each structure. **(3 marks)**

Plant and algal cells contain a permanent **vacuole**. This is filled with cell sap and helps to keep the cell rigid.

They have a **cell wall** made of cellulose, which strengthens the cell.

They also contain **chloroplasts**. These are the structures where photosynthesis takes place to make food for the cell.

Algae are simple aquatic organisms. Their cells have a similar structure to plant cells, so algae used to be classified as plants. They are now classified as protists (see pages 36 and 95).

Take care not to confuse the cell wall with the cell membrane.

Bacterial cells also have cell walls, but these are not made from cellulose.

Now try this

1 Name the cell structures that are found in both animal cells and plant cells. **(5 marks)**

2 Muscle cells contain more mitochondria than skin cells. Suggest an explanation for this. **(2 marks)**

3 Cells found in plant roots do not usually contain chloroplasts. Suggest an explanation for this. **(2 marks)**

Eukaryotes and prokaryotes

Animal cells and plant cells are **eukaryotic** cells, but bacterial cells are **prokaryotic** cells.

Bacterial cells

Eukaryotic cells and prokaryotic cells have:
- a cell membrane
- cytoplasm.

But prokaryotic cells:
- are surrounded by a cell wall
- do not have a ~~nucleus~~.

A single loop of **chromosomal DNA** lies free in the cytoplasm. This carries most of the bacterial genes.

cell membrane

Some bacteria have a **flagellum** to help them move.

Ribosomes are tiny structures that make proteins.

Some bacteria have extra circles of DNA called **plasmid DNA.** Plasmids contain additional genes that are not found in chromosomes.

Many bacteria have a **cell wall** for protection, but it is made of different substances to plant cell walls.

Standard form

Numbers in standard form have two parts.

$$7.3 \times 10^{-6}$$

This part is a number greater than or equal to 1 and less than 10.

This part is a power of 10.

You can use standard form to write very large or very small numbers.

$$920\,000 = 9.2 \times 10^5$$

Numbers greater than 10 have a positive power of 10.

$$0.007\,03 = 7.03 \times 10^{-3}$$

Numbers less than 1 have a negative power of 10.

🖩 Maths skills Finding a power

You can count decimal places to convert between numbers in standard form and ordinary numbers.

3 jumps

$$7\,9\,0\,0 = 7.9 \times 10^3$$

7900 > 10 so the power is positive

4 jumps

$$0.0\,0\,0\,3\,5 = 3.5 \times 10^{-4}$$

0.00035 < 1 so the power is negative

Be careful!

Don't just count zeros to work out the power.

Worked example

An animal cell is 0.01 mm long and a bacterial cell is 1 μm long.

(a) Give these lengths in metres, m, in standard form.

(2 marks)

$0.01\,mm = 0.01 \times 10^{-3}\,m = 1 \times 10^{-5}\,m$

$1\,μm = 1 \times 10^{-6}\,m$

(b) How many orders of magnitude greater is the size of the animal cell?   **(1 mark)**

$0.01\,mm = 0.01 \times 10^3\,μm = 10\,μm$

So the animal cell is $\dfrac{10}{1} = 10$ times larger. This is one order of magnitude.

These units use the prefixes milli, m, and micro, μ. They refer to powers of 10:

milli-	micro-	nano-
10^{-3}	10^{-6}	10^{-9}

1 nm (nanometre) is $1 \times 10^{-9}\,m$.

You can also work this out using the powers of 10 in the lengths given in standard form.

The difference between 5 and 6 is 1, showing one order of magnitude difference.

Now try this

(a) Give these lengths in metres, m, in standard form:
    (i) 75 μm (the diameter of a plant cell); (ii) 750 nm (the diameter of a mitochondrion).   **(2 marks)**
(b) Determine how many orders of magnitude greater a plant cell is than a mitochondrion.   **(1 mark)**

3

Specialised animal cells

Animal cells may be specialised to carry out a particular function.

Ciliated cells

Cilia are tiny hair-like structures on the surface of some cells. They can move in organised ways to push substances past the surface of the cell.

cilia
ribosomes
mitochondrion
cell membrane
cytoplasm
nucleus

Ciliated cells line the **trachea** or windpipe. These sweep mucus away from the lungs, carrying away dirt and bacteria.

Differentiation

As an organism develops, cells **differentiate** to form different specialised cells:

☑ Most types of animal cell differentiate at an early stage.

☑ Cells acquire different structures so they can carry out a particular function.

☑ Cell division in mature animals is mainly for repair and replacement.

You can revise the structures in generalised animal cells on page 2 and stem cells on page 10.

Sperm cells   Haploid

Sperm cells carry the male parent's genetic information to the female parent's egg cell.

They may have to travel a relatively long way, then get through the surface of the egg cell.

nucleus
acrosome
tail

A sperm cell has:
- a tail, and many mitochondria to release the energy needed for the tail to move movement
- enzymes in its acrosome to digest the outer layers of an egg cell.

Nerve cells

Nerve cells carry electrical impulses from one part of the body to another, which may be a long way apart. Nerve cells often connect with one another.

dendrite
nucleus
axon

A typical nerve cell has:
- a long axon to connect distant parts
- tiny finger-like dendrites that make connections with other nerve cells.

Worked example

The diagram shows muscle cells. Describe, using information shown, how they are specialised for their function. **(2 marks)**

nucleus
protein fibres

The protein fibres slide over each other, making the muscle cells contract.

Muscle cells also contain many mitochondria. These supply the energy needed to keep the muscles working.

Now try this

Different specialised cells have different shapes. Explain how the shapes of ciliated cells in the trachea and sperm cells are related to their functions.
(4 marks)

Specialised plant cells

Plant cells may be **specialised** to carry out a particular function.

Root hair cells

Root hair cells are found on the surface of plant roots. They absorb water and mineral ions.

- nucleus
- vacuole
- cell wall
- cytoplasm
- cell membrane
- soil particle

A root hair cell has: *maintain con gradient*

- a large vacuole to increase the rate of absorption of water by **osmosis**
- many mitochondria to release the energy needed for the **active transport** of mineral ions.

> You can revise diffusion on page 11, osmosis on page 13 and active transport on page 15.

Differentiation

As an organism develops, cells **differentiate** to form different specialised cells:

☑ Meristem cells in the tips of roots and shoots are undifferentiated cells.

☑ Cells acquire different structures so they can carry out a particular function.

Unlike most types of animal cells, many types of plant cells keep the ability to differentiate through their life.

You can revise the structures in generalised plant cells on page 2 and stem cells on page 10.

Xylem cells

Xylem cells form the **transport tissue** that carries water and mineral ions from roots to the rest of the plant. Lignin builds up in xylem cells. These die and form hollow tubes.

(4 marks)

- *xylem cells are form*
- *lignin build up in cell wall*
- *the xylem cell dies*
- *the end wall is broken down forming a long tube*

The wall that makes the tube is made of lignin.

Xylem cells have:

- **lignin** to provide strength and support
- no end walls, so water and mineral ions can flow easily through the plant
- **pits** that let water and ions move out.

Worked example

Phloem cells form the transport tissue that carries dissolved sugars through a plant.

- sieve plate
- *sieve plate*
- companion cell
- sieve tube
- *mitochondria*
- cytoplasm

Explain two ways in which phloem cells are specialised for their function. **(4 marks)**

The cell walls between cells form sieve plates, which let solutions move from cell to cell. They have companion cells which contain lots of mitochondria. These release the energy needed for the active transport of sugars.

Now try this

Root hair cells have large surface areas to increase the rate of absorption of water and mineral ions. Describe the structure that provides this feature. **(2 marks)**

> Mention the shape of this structure and where it grows.

Using a light microscope

 Practical skills You can use a light microscope to observe, draw and label a selection of plant and animal cells.

Core practical

Observing cells

Aim

to make a slide with onion cells, then use a microscope to observe, draw and label these and other cells

Apparatus

- small piece of onion
- forceps
- microscope slide with coverslip
- iodine solution
- prepared slides of animal and plant cells
- light microscope

Method: Preparing a specimen

1. Use the forceps to peel a thin layer of epidermal tissue from the onion.
2. Carefully lay the thin layer on the slide.
3. Add one or two drops of iodine solution.
4. Lower the coverslip over the onion layer. Remove excess liquid with tissue paper.

Method: Observing a specimen

5. Place a slide on the microscope stage.
6. Turn the turret to the lowest power objective lens.
7. Looking from the side, use the coarse wheel to lower the lens so it almost touches the slide.
8. Looking through the eyepiece, use the coarse wheel to raise the objective lens until the image is focused.
9. Turn the turret to select a higher power objective lens. Use the fine wheel to focus the image again.
10. Repeat steps 5–9 with other slides.

Results

Make a pencil drawing of one or more cells:

- Draw outlines of what you see.
- Use shading only to distinguish between structures.
- Identify structures using labels with straight lines.
- Write the magnification on your drawing.

> Onion skin is easily peeled away in single layers of cells.

> Wear eye protection when you handle the iodine solution. Wash off spills straight away because it will stain skin and clothes.

> You may find it easier to position the layer of onion if you put a drop of water on the slide first. Don't let the tissue curl over on itself.

> Place one edge of the coverslip on the slide at 45°. Gently lower the other edge so that liquid and bubbles are pushed out.

> You can revise calculating magnification on page 1.

The light microscope

- eyepiece
- objective
- stage with clips to hold slide
- coarse focusing wheel
- fine focusing wheel
- mirror to reflect light through slide

Some modern microscopes have a lamp rather than a mirror.

Now try this

Make a labelled drawing of the onion epidermal cells.
Identify the cell wall, nucleus, cytoplasm and a chloroplast. **(5 marks)**

Aseptic techniques

Bacteria can be grown in a nutrient broth solution or as colonies on an agar gel plate.

Sterilising apparatus and media

Killing microorganisms is called **sterilisation**.

For use in the lab:
* **sterile** Petri dishes, <u>irradiated</u> with gamma radiation at the factory, can be used
* culture media (such as nutrient broth or agar gel) can be sterilised using an **autoclave**, which uses steam at high temperature under pressure
* apparatus can be kept sterile during working by using a flame.

Binary fission

Bacteria reproduce by simple cell division, called **binary fission**.

Bacteria can divide as often as once every 20 minutes if:
* they have enough **nutrients**, and
* they are at a suitable **temperature**.

When investigating the growth of bacteria, it is important to use uncontaminated cultures, and sterile apparatus and culture media.

Sterilising apparatus

A wire loop is used to transfer a sample of a bacterial culture from one place to another.

The loop is sterilised in a hot flame and then cooled, before using it to transfer microorganisms to the culture medium.

Human body temperature is 37 °C. In school laboratories, cultures should generally be incubated at 25 °C. This is to reduce the risk of harmful bacteria growing.

Incubating Petri dishes

The layer of sterile agar gel on a Petri dish is inoculated with bacteria using a wire loop:
* The Petri dish is opened upside down near a flame.
* The loop is swept over the agar gel.
* The lid is replaced and sealed.
* The dish is labelled with the worker's name, the date, and what it contains.

The dish is not sealed all round. This is to stop the growth of harmful anaerobic bacteria.

Worked example

The mean division time for a strain of bacteria is 0.5 hours. A culture with 1000 of these bacteria is incubated. Calculate the number of bacteria after 4 hours. **(3 marks)**

number of divisions = $\frac{4}{0.5}$ = 8

number of bacteria = 1000 × 2^8

= 1000 × 256

= 256 000

Maths skills — Using powers

Calculators usually have a button marked $\boxed{x^y}$

To calculate 2^8 (two to the power eight) on your calculator, enter: 2 $\boxed{x^y}$ 8 $\boxed{=}$

Some calculators have a button marked $\boxed{y^x}$

If neither button is present, instead enter:

2 × 2 × 2 × 2 × 2 × 2 × 2 × 2

Now try this

1 Describe **three** ways in which apparatus and culture media can be sterilised. **(3 marks)**
2 Suggest an explanation for why the laboratory bench is wiped with an antiseptic solution before working with bacteria. **(2 marks)**
3 The mean division time for a strain of bacteria is 40 minutes. A culture of 2500 of these bacteria is incubated. Calculate the number of bacteria after 4 hours. **(2 marks)**

An antiseptic is a substance that kills microorganisms or prevents their growth.

Investigating microbial cultures

 Practical skills You can measure the effect of antiseptics and antibiotics by measuring **zones of inhibition.**

Core practical

The effects of antiseptics

Aim

to investigate the effect of antiseptics on bacterial growth using agar plates

Apparatus

Petri dish with bacteria spread over nutrient agar • filter paper discs, about 4 mm in diameter • three different antiseptics • forceps • adhesive tape • permanent marker pen • Bunsen burner and heat-resistant mat

Method

1. Turn the Petri dish upside down and mark the bottom into three equal segments.
2. Mark a spot in the middle of each segment. Number each segment. Write your name and the date on the bottom.
3. Add a different antiseptic to each disc. Record what you added to each segment.
4. Partly lift the Petri dish and use the forceps to place a disc over each spot.
5. Use adhesive tape to secure the lid.
6. Incubate the plate for 2 days at 25 °C.

Results

Measure the diameter of the clear zone around each paper disc:

- Measure in millimetres, mm.
- Make two measurements at 90° to each other, and calculate the mean diameter.
- Calculate the area of each clear zone, giving your answers in standard form in metres squared, m².

You could also investigate the effects of antibiotics on bacterial growth.

Your teacher or school technician will prepare the Petri dish for you. They will:
- add a few drops of a bacterial culture to the agar on the Petri dish
- use a sterile glass spreader to spread the bacteria over the agar.

Certain strains of *E. coli* are suitable to use.

Make sure you wipe the bench with a disinfectant solution, then dry it with clean paper towels before starting.

Before you use the forceps, briefly pass the end of the forceps through a Bunsen burner flame to sterilise them.

 Maths skills **The area of a circle**

You can calculate the area of a circle using:

$$area = \pi r^2$$

where r = radius.

Divide the diameter by 2 to find the radius.

Remember that $1\,mm = 1 \times 10^{-3}\,m$.

Now try this

A student investigates the effect of three different antiseptics on the growth of bacteria. The table shows her results.

| Antiseptic | Diameter of the clear zone (mm) | | Mean value | Area (m²) |
	Measurement 1	Measurement 2		
A	14	16	15	1.77×10^{-4}
B	24	26		
C	19	21		

(a) Calculate the mean diameter and mean area for antiseptics B and C. **(4 marks)**

(b) Explain what the results show. **(2 marks)**

Mitosis

The nucleus of a cell contains **chromosomes** made of **DNA** molecules which carry a large number of **genes**. The chromosomes are normally found in pairs in body cells.

The cell cycle

There are three main stages in the **cell cycle**:

 interphase, the stage where the cell:
- grows
- increases the numbers of structures such as ribosomes and mitochondria
- **replicates** its DNA to form two copies of each chromosome.

② **mitosis**, the stage where:
- <u>one set</u> of chromosomes is <u>pulled to each end of the cell</u>
- <u>the nucleus divides.</u>

③ **cell division**, the stage immediately after mitosis where:
- cell membranes and cytoplasm divide
- two identical cells form.

Interphase

A cell spends most of its life in interphase.

interphase — mitosis and cell division

Three things to remember

☑ mi-To-sis makes Two cells

☑ mi-tosis makes cells that are genetically Identical to the parent cell

☑ Diploid means Double (two sets of) chromosomes

Parent cell and two daughter cells

The cell that divides is called the **parent cell**.

The parent cell divides to form two **daughter cells**.

The parent cell is a **diploid** cell. This means it has two sets of chromosomes.

Before the parent cell divides, each chromosome is copied exactly.

When the cell divides in two, each cell gets one copy of each chromosome.

nucleus

The chromosomes are drawn short here, and coloured, so it is easier to see what is happening. They don't really look like this.

The daughter cells are genetically identical to each other and the parent cells. They are also diploid cells.

You can revise asexual reproduction on page 77.

Worked example

Give three situations in which mitosis is used to produce new cells. **(3 marks)**

Mitosis produces new body cells for:
- growth
- repair to damaged parts of the body.

It also produces new individuals by asexual reproduction.

Cell division by mitosis is important in the growth and development of multicellular organisms.

Now try this

Explain why daughter cells produced by mitosis are genetically identical to each other and to the parent cell.

(3 marks)

9

Stem cells

Stem cells are undifferentiated cells in an organism. They can produce many more cells of the same type, and specialised cells by **differentiation**.

Stem cells in animals

A zygote is a single cell formed when an egg cell and sperm cell join together. This divides to form a ball of cells called an **embryo**:
- the embryo contains **embryonic stem cells**, which differentiate to form many different specialised cell types.

Adult animals retain some stem cells. The bone marrow contains **adult stem cells**. These form many cell types, including blood cells.

Stem cells in plants

The **meristems** in plants contain stem cells.

Meristem tissue is found at the tips of roots and shoots. It differentiates throughout the life of the plant:
- in roots to form root hair cells, xylem, phloem and other cells in the root
- in shoots to form mesophyll cells, xylem, phloem and other cells in the stems and leaves.

Stem cell treatment

Stem cells may be able to help to treat conditions such as diabetes and paralysis. Injected stem cells may be able to replace damaged or missing cells in the pancreas or nervous system.

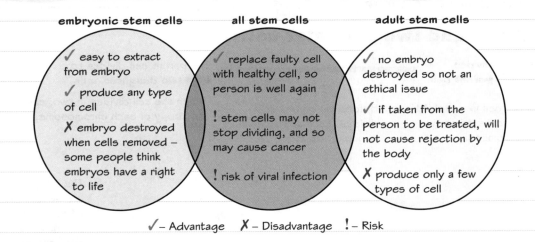

embryonic stem cells

✓ easy to extract from embryo

✓ produce any type of cell

✗ embryo destroyed when cells removed – some people think embryos have a right to life

all stem cells

✓ replace faulty cell with healthy cell, so person is well again

! stem cells may not stop dividing, and so may cause cancer

! risk of viral infection

adult stem cells

✓ no embryo destroyed so not an ethical issue

✓ if taken from the person to be treated, will not cause rejection by the body

✗ produce only a few types of cell

✓ – Advantage   ✗ – Disadvantage   ! – Risk

Worked example

Stem cells from meristems in plants can be used to produce clones rapidly and economically. Explain one possible use of producing plants this way. **(2 marks)**

If a crop plant has a useful feature, such as resistance to disease, it could be cloned. This would provide large numbers of these plants for farmers to grow.

Clones are genetically identical individuals.

Cloning could also save rare plant species from extinction by producing large numbers of these plants.

You can revise cloning in plants on page 88.

Now try this

Therapeutic cloning is a way to produce an embryo using a cell from a patient. It involves replacing the nucleus in an egg cell with the nucleus from one of the patient's cells. The egg cell is then stimulated to make it divide to form an embryo.
Explain an advantage and a disadvantage of treating a patient with stem cells produced in this way.   **(4 marks)**

Diffusion

Substances may move into and out of cells by **diffusion** across cell membranes.

Net movement of particles

Diffusion is the spreading out of:
- the particles of a gas, or
- the particles of any substance in solution.

Diffusion results in an overall movement of particles down a **concentration gradient**:
- from an area of <u>higher concentration</u> to an area of <u>lower concentration</u>.

The rate of diffusion increases when:
- the <u>difference in concentration increases</u>, so the concentration gradient increases
- the <u>temperature increases</u>, so the particles move more quickly.

The number of particles decreases as you go down a concentration gradient.

diffusion

higher concentration   lower concentration

Diffusion explains why you can smell cooking from the next room, and why ink spreads through a glass of water without stirring.

Without using any energy!

Diffusion across membranes

Diffusion also happens across membranes.

In addition to the two factors above, the rate of diffusion <u>increases</u> when:
- the **surface area** of the membrane **increases**
- the **thickness** of the membrane **decreases**.

Molecules pass through the membrane in both directions. The rate of movement from left to right is greater than the rate of movement from right to left.

There is a net movement of particles from left to right until their concentration is the same on both sides.

Particles remain in random and continuous motion before, during and after diffusion.

Worked example

Describe three examples of the movement of substances into or out of animal cells by diffusion. **(3 marks)**

During gas exchange, oxygen moves into cells by diffusion, and carbon dioxide moves out of the cells by diffusion.

Urea is a waste product. It moves out of cells into the blood plasma by diffusion.

Diffusion also happens in plants, such as gas exchange in the leaves during photosynthesis.

You can revise gas exchange in the lungs on page 24, and the function of the kidneys on pages 64 and 65.

Now try this

1 Explain diffusion in terms of the particles involved.   **(3 marks)**
2 Explain how the rate of diffusion changes as the temperature increases.   **(2 marks)**

Exchange surfaces

Multicellular organisms have adaptations to allow effective exchange of materials.

Surface area to volume ratio

Single-celled organisms have high **surface area to volume ratios**. This means:

👍 there is a relatively large surface area of cell membrane for diffusion to occur

👍 the distances from the cell membrane to the interior of the cell are small.

Molecules move in and out of a single-celled organism well enough to meet its needs.

However, multicellular organisms need:

- **exchange surfaces** to provide large surface area to volume ratios
- **transport systems** to maintain high concentration gradients.

🖩 Maths skills — Calculating surface area to volume ratio

Imagine a cube-shaped cell with sides 1 μm long. It has six equal square sides:
- surface area = 6 × (1 × 1) = 6 μm²
- volume = 1 × 1 × 1 = 1 μm³

The cell's surface area to volume ratio is: $\frac{6}{1} = 6$

Surface area to volume ratio decreases with increasing size. For example, for cubes:

Length of side (μm)	Surface area to volume ratio
10	0.6
100	0.06

Increasing the effectiveness of an exchange surface

The effectiveness of an exchange surface increases with:

- increase in surface area
  - e.g. alveoli in lungs, or villi in small intestine
- a shorter distance for diffusion
  - e.g. surfaces are one cell thick
- maintenance of a high concentration gradient
  - e.g. animals have an efficient blood supply, lungs are ventilated

The fast removal of substances maintains a <u>steep concentration gradient</u> so diffusion is faster.

\# ออกข้อ

For example, the lining of the small intestine is folded into many tiny <u>finger-like</u> structures called (villi.) These provide:
- a large surface area for diffusion
- a <u>single layer of cells</u> for a short diffusion path
- a network of capillaries to carry food molecules away.

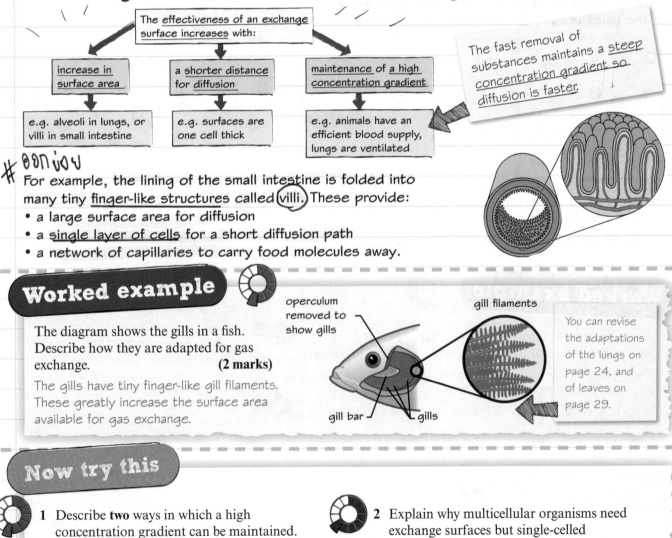

Worked example

The diagram shows the gills in a fish. Describe how they are adapted for gas exchange. **(2 marks)**

The gills have tiny finger-like gill filaments. These greatly increase the surface area available for gas exchange.

operculum removed to show gills

gill filaments

gill bar — gills

You can revise the adaptations of the lungs on page 24, and of leaves on page 29.

Now try this

1. Describe **two** ways in which a high concentration gradient can be maintained. **(2 marks)**

2. Explain why multicellular organisms need exchange surfaces but single-celled organisms do not. **(4 marks)**

Osmosis

Water may move across cell membranes by **osmosis**.

Explaining osmosis

Osmosis is the diffusion of water:
- from a dilute solution
- to a concentrated solution
- through a (**partially permeable membrane.**)

A partially permeable membrane:
- 👍 allows small molecules such as water molecules to pass through it
- 👎 does not allow ions or larger molecules such as sucrose to pass through it.

Cell membranes are partially permeable. Water can diffuse across them but sugars cannot.

This diagram shows how there is a net movement of water from a dilute sucrose solution to a concentrated sucrose solution.

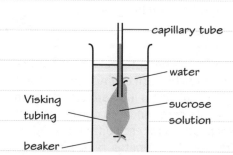

Investigating osmosis

Visking tubing is an artificial partially permeable membrane. It is supplied dry and is soaked in water to soften it before use.

The diagram on the right shows a typical osmosis experiment. The Visking tubing is:
- tied at the bottom with cotton thread
- filled with sucrose solution
- tightly fitted around a narrow glass capillary tube and tied with thread
- placed in a beaker of water.

There is a net movement of water from the beaker into the sucrose solution.

Over time, the volume of liquid inside the Visking tube increases. The level of liquid in the capillary tube rises.

Worked example

> You see similar results with other types of cell. Cells increase in size, and may even burst, if placed in water or very dilute solutions.

Red blood cells are placed on a microscope slide. A few drops of concentrated sucrose solution are added, and the red blood cells are observed using a microscope. They decrease in size. Explain these observations. **(2 marks)**

The sucrose solution is more concentrated than the cytoplasm in the red blood cells. Water moves across the cell membrane from the cytoplasm into the sucrose solution, causing the red blood cells to shrink.

> This is an example of osmosis, with the cell membrane acting as a partially permeable membrane.
>
> You may see the term 'semi-permeable membrane' instead of 'partially permeable membrane'.

Now try this

1 Describe what is meant by osmosis. **(3 marks)**

2 Sugar is sprinkled over some cut pieces of fruit. A liquid gradually appears on the cut surfaces of the fruit. Explain this observation. **(3 marks)**

> The sugar dissolves to form a concentrated solution.

Investigating osmosis

 You can investigate the effect of **osmosis** on the mass of plant tissue.

Core practical

Concentration and osmosis

Aim

to investigate the effect of different sucrose concentrations on the mass of potato tissue

> You could use different concentrations of sodium chloride instead.

Apparatus

- potato
- cork borer, cutting board and knife
- ruler with 1 mm scale divisions
- balance, precise to ±0.01 g
- different concentrations of sucrose
- boiling tubes with bungs and labels

> A suitable range is 0–1.0 mol/dm³ sucrose solution. Distilled water provides the 0 mol/dm³ solution.
> If you use sodium chloride, use 0–0.5 mol/dm³ (0–1.0 mol/dm³ dissolved ions).

> You could use labelled Petri dishes instead.

Method

1. Use the cork borer to make potato cylinders. Trim them to about 30 mm.
2. Two-thirds fill each boiling tube with a different sucrose solution.
3. Dry the outside of a potato cylinder with a paper towel, then measure and record its mass. Add it to one of the solutions.
4. Repeat step 3 for each different solution.
5. After about 30 minutes, remove each potato cylinder. Dry them and record their mass.

> Make sure all the potato cylinders are the same length.

> You could also measure and record the new lengths, and then analyse those results.

Results

> You could wait for up to 24 hours instead.

Make a suitable table to record your results. This one has some example results in it.

Sucrose conc. (mol/dm³)	Start mass (g)	End mass (g)	Change (g)	% change
0	2.00	2.53	+0.53	+26.5
0.25	2.15	2.23		
0.50	2.05	1.80		
0.75	1.96	1.53		
1.00	2.08	1.45	−0.63	−30.3

Analysis

1. Calculate the change in mass and percentage change in mass for each potato cylinder.
2. Plot a graph with:
   - percentage change in mass on the vertical axis
   - concentration of sucrose on the horizontal axis.

🔢 Maths skills Calculating percentage change

percentage change in mass =

$$\frac{(\text{final mass} - \text{initial mass})}{\text{initial mass}} \times 100$$

A negative value means the mass decreased.

Now try this

(a) Calculate the percentage change in mass for the results in the table. **(3 marks)**
(b) Plot a suitable graph using all the percentage changes in mass. **(3 marks)**
(c) Use your graph to estimate the concentration of the potato tissue. **(1 mark)**

> This is when the percentage change in mass is zero.

Active transport

Unlike diffusion and osmosis, **active transport** requires **energy** from respiration to happen.

Explaining active transport

Active transport moves substances:
- from a more dilute solution
- to a more concentrated solution, and
- this <u>requires energy</u> from **respiration**.

Active transport lets substances move (against) a **concentration gradient**.

Cell membranes contain 'transport proteins'. The diagram shows how energy allows one of these molecules to change shape, which moves a solute particle into a cell.

outside the cell   cell membrane

+ energy

inside the cell   transport protein

Active transport in the gut

Glucose is provided by the food we eat. It must be absorbed across the wall of the **small intestine** into the bloodstream. Active transport of glucose molecules means:
- glucose can be absorbed, even if its concentration in the blood is higher than its concentration in the small intestine.

It is important because cells need glucose for respiration. This happens in the mitochondria (you can revise cell structure on page 2).

Transport processes   Note!!

Process	Features
diffusion	• Substances move from high concentration of substance to low concentration. • Membranes are not needed. • No energy is needed.
osmosis	• Water moves from a dilute solution to a concentrated solution. • Partially permeable ✳️ membrane is needed. • No energy is needed.
active transport	• Substances move from low concentration of substance to high concentration. • Membranes are needed. ✳️ • Energy is needed. ✳️

Active transport in roots

Plants need **mineral ions** such as nitrate ions, NO_3^-, for healthy growth. The **root hair cells** absorb mineral ions from soil water. The concentration of ions in soil water is very low. Active transport of mineral ions means:
- plants can absorb mineral ions, even from very dilute solutions.

(you can revise cell structure on page 2).

Worked example

A farmer's field becomes flooded with water after heavy rain. Explain why the plants growing there are unable to absorb enough nitrate ions for healthy growth.   **(4 marks)**

Plants absorb nitrate ions from the soil by active transport. This needs energy from cell respiration. Water-logged soil contains very little air. There is not sufficient oxygen for enough active transport by root hair cells.

Plants need nitrate ions so they can make proteins. Nitrate deficiency causes stunted growth in plants.

Some substances are toxic to cells because they prevent respiration. Active transport will also stop in the presence of these substances.

Now try this

Explain why cells involved in active transport contain many mitochondria.   **(2 marks)**

Extended response – Cell biology

There will be at least one 6-mark question on your exam paper. For these questions, you will need to think scientifically and structure your answer logically, showing how the points you make are related to each other.

For the questions on this page, you can revise cells and transport on pages 2 and 11–15.

Worked example

Substances are transported into and out of cells by diffusion, osmosis and active transport. Compare these three processes. In your answer, you should include a suitable example of each process in living organisms. **(6 marks)**

Diffusion is the spreading out of particles, producing a net movement from an area of higher concentration to an area of lower concentration. It is important for gas exchange in the lungs and gills.

Osmosis also involves diffusion, but only of water. It is important for water absorption by root hair cells. Unlike diffusion, which can take place in gases or in solutions, osmosis only involves solutions.

In addition, while diffusion can happen across cell membranes but does not have to, osmosis only happens across partially permeable membranes such as cell membranes.

Active transport is important for the absorption of mineral ions by plant root hair cells. During active transport, substances move from a more dilute solution to a more concentrated solution. This is against a concentration gradient, whereas diffusion happens down a concentration gradient.

Diffusion and osmosis are passive processes that do not need energy from respiration. However, active transport does need energy from respiration and cannot happen without it.

Command word: Compare

When you are asked to **compare** something, you must describe the similarities and/or differences between two or more things. It is not enough just to write about one of the things.

This part of the answer defines diffusion and gives suitable examples, but does not yet compare it with the other two processes. Carbon dioxide and oxygen could be named as the gases involved.

The answer should have stated that osmosis involves the net movement of water from a dilute to a more concentrated solution.

The use of the connective 'in addition' signals that a related idea is about to follow.

The use of the connective 'whereas' signals that an opposite idea is about to follow.

The use of the connective 'however' also signals that an opposite idea follows.

You need to show comprehensive knowledge and understanding, using relevant scientific ideas, to support your answer. The three transport processes are defined, and suitable examples of each one are given. Their similarities and differences are identified in a logical way.

Now try this

Cells are the basic unit of all animals and plants. Compare the structure of a typical animal cell with the structure of a typical plant cell. In your answer, you should name the main sub-cellular structures and describe their functions. **(6 marks)**

You must provide a written answer, but a labelled diagram can help to make some points clearer.

The digestive system

An **enzyme** is a biological catalyst: a protein that can speed up or control a specific reaction.

Digestive enzymes

Digestive enzymes convert larger food molecules into smaller, soluble molecules that can be absorbed into the bloodstream:

 Carbohydrase enzymes

carbohydrates → simple sugars

Amylase is produced by the salivary glands, pancreas and small intestine. It breaks down **starch** in the mouth and small intestine.

 Protease enzymes

proteins → amino acids

Produced by the stomach, pancreas and small intestine. Breakdown of **proteins** occurs in the stomach and small intestine.

③ **Lipase** enzymes

lipids → fatty acids + glycerol

Produced by the pancreas and small intestine. Breakdown of **lipids** (fats and oils) occurs in the small intestine.

The digestive system

salivary glands
produce digestive enzymes

liver
produces bile

small intestine
digestion completed and soluble substances absorbed

stomach
some digestion happens here

pancreas
produces digestive enzymes

large intestine
water absorbed and faeces formed

The **digestive system** is an organ system in which several organs work together to digest and absorb food.

Worked example

Describe two uses of the products of digestion by the body. **(2 marks)**

Simple sugars can be used to build new carbohydrates.

Some of the glucose produced by digestion of carbohydrates is used in respiration.

Other products of digestion can be used to build other new substances:
• amino acids for new proteins
• fatty acids and glycerol for new lipids.

You can revise the role of enzymes in metabolism (all the reactions in a cell or body) on page 51.

Principles of organisation

The bodies of large multicellular organisms are organised into different **organ systems**.

Cell	Tissue	Organ	Organ system
basic building block of all living organisms	group of cells with a similar structure and function	combination of tissues, each with specific functions	two or more organs working together
e.g. muscle cell, secretory cell	e.g. muscular tissue, glandular tissue, epithelial tissue	e.g. heart, stomach, lungs, kidney	e.g. digestive system, circulatory system, respiratory system

Now try this

Make a table to summarise the features of amylase, proteases and lipases. In your table, show their sites of production and action, the substances they break down and the products formed. **(6 marks)**

Food testing

Practical skills You can use **qualitative reagents** to test for a range of carbohydrates, lipids and proteins.

Core practical

Chemical tests for food

These are **qualitative tests**. They show if a substance is present but not how much.

Aim

to investigate different types of food to identify the substances they contain

Apparatus

- eye protection
- test tubes and test-tube rack
- hot water bath
- distilled water
- Benedict's solution
- iodine solution
- Sudan III stain in ethanol solution
- Biuret reagent
- food samples

Make a hot water bath by half filling a beaker with hot (not boiling) water from a kettle.

Wash spills off clothing or skin immediately.

Ethanol is highly flammable.

If you are given larger pieces of food, cut them into small pieces or grind them with a pestle and mortar.

Methods

Before each test, mix a small food sample with about 2 cm³ of distilled water in a test tube.

Test for reducing sugars: Add a few drops of Benedict's solution and heat in the hot water bath for 5 minutes. Record any colour change.

Test for starch: Add a few drops of iodine solution. Record any colour change.

Test for lipids: Add a few drops of Sudan III stain solution and shake to mix. Record whether a red upper layer forms.

Test for proteins: Add an equal volume of Biuret reagent and shake to mix. Record any colour change.

Glucose and fructose are **reducing sugars**, but sucrose (table sugar) is not.

You could just use ethanol instead, and look for a cloudy upper layer.

Results

Make a table to record your results. This one shows the expected results for each substance.

Test reagent	Substance detected	Colour of reagent or negative result	Colour or result if substance present
Benedict's solution	reducing sugars	blue	green → orange → red (depending on amount of sugar)
iodine solution	starch	orange-brown	blue-black
Sudan III stain in ethanol	lipids	colourless upper layer	red upper layer
Biuret reagent	proteins	blue	pink or purple

important

Now try this

Suggest **two** reasons that explain why a hot water bath is safer to use in this investigation, rather than a Bunsen burner. **(3 marks)**

Think about how the test for reducing sugars is carried out, and the nature of the substance used in the test for lipids.

Enzymes

Enzymes are proteins, they can be denatured by temperature.

The **lock and key theory** is a simple model to explain the action of enzymes.

Lock and key theory

Enzymes are protein molecules, with complex shapes which are important for their activity:
- Part of an enzyme molecule is its **active site**.
- A **substrate** molecule fits into the active site, like a key fits into a lock.
- Other substances do not fit, so are not substrates for the enzyme.
- If the shape of an active site changes, the enzyme is **denatured** and no longer works.

active site

enzyme molecule

substrate molecule

substrate molecule fits the active site

reaction

products released

Effect of temperature

The rate of chemical reactions increases as the **temperature** increases. This happens with enzyme-catalysed reactions, but:
- There is an **optimum temperature** at which there is a maximum reaction rate.
- At higher temperatures the active site changes shape, reducing the reaction rate.

Optimum temperature
The enzyme works at its maximum rate.

Heating begins to change the shape of the active site.

Rate increases as the rate of collisions between the substrate active site increases.

Enzyme is **denatured**.

Rate of reaction

Temperature in °C
10 20 30 40 50 60

Effect of pH

The **pH** of an enzyme's surroundings affects the shape of the active site. This means:
- Enzymes have an **optimum pH** at which there is a maximum reaction rate.
- Below and above the optimum pH, the bonds in the enzyme are affected, so the shape of the active site changes.

Optimum pH (different for different enzymes)

Enzyme is **denatured**.

Enzyme is **denatured**.

Rate of reaction

pH
4 5 6 7 8 9

Worked example

Bile is made in the liver, stored in the gall bladder and released into the small intestine. Explain its role in the digestion of fats.   **(3 marks)**

Bile is alkaline, so it neutralises hydrochloric acid made by the stomach. It also emulsifies fat, breaking it into small droplets. This increases the surface area of the fat. The alkaline conditions and the large surface area increase the rate at which lipases break down fat.

Partly digested food is acidic as it reaches the small intestine. The optimum pH for the lipase is pH 7.5–8.0 (alkaline conditions).

Now try this

1 Proteases digest proteins in the stomach. Suggest the optimum pH of these proteases, and justify your answer.   **(2 marks)**

2 Explain the importance of the active site for enzyme activity at different temperatures and pH values.   **(4 marks)**

Investigating enzymes

 Practical skills You can use **continuous sampling** to find the time taken to digest starch at different pH values.

Core practical

Investigating the effect of pH

Aim

to investigate the effect of pH on amylase activity

Apparatus and Method

1 Add one drop of iodine solution to each well in a spotting tile.
2 Add the following to a test tube: 2 cm³ of a pH buffer solution, 2 cm³ of amylase solution, and 2 cm³ of starch solution.
3 Mix with a glass rod and start a stop clock.
4 After 30 seconds, use a glass rod to transfer one drop to a well in the spotting tile. Rinse the glass rod with water.
5 Repeat step 4 until there is no change in the colour of the iodine solution. Stop the stop clock and record the time.
6 Repeat steps 1 to 5 using different pH buffer solutions.

> Amylase is an enzyme that breaks down starch into simple sugars.

> All the solutions should be kept at the same temperature. You can achieve this by keeping the bottles of solution in a water bath maintained at 25°C.

> When the iodine in a well no longer changes colour, there is no starch present – it has all been digested to form simple sugars.

> A **buffer solution** is a mixture that resists changes to its pH.

Results

Record your results in a suitable table. This one shows some sample results.

pH of buffer	Time taken to digest starch (s)	Rate of reaction (/s)
4	169	0.0059
5	125	
6	100	
7	118	
8	260	

Analysis

1 Calculate the rate of reaction that corresponds to each reaction time.
2 Plot a graph with:
  • rate of reaction on the vertical axis
  • pH of buffer solution on the horizontal axis.
3 Use your results to identify the **optimum pH** for amylase activity.

> The optimum pH is the pH that gives the greatest rate of reaction.

Maths skills ### Rate of reaction

The rate of a reaction is proportional to the time taken for it to finish. This means you can convert reaction times to reaction rates:

$$\text{rate (/s)} = \frac{1}{\text{time (s)}}$$

For example, if reaction time is 200 s:
• reaction rate = 1/200 = 0.005 /s

> Choose sensible scales that allow the plotted points to occupy at least 50% of the area of the graph. Draw a curve of best fit.

Now try this

1 Calculate the rates of reaction for the reaction times shown in the table. Give your answers to 2 significant figures. **(2 marks)**
2 Draw a suitable graph to show the rates of reaction between pH 4 and 8. **(3 marks)**
3 Use your graph drawn for question 2, or the table of results, to identify the optimum pH for amylase. Explain your answer. **(2 marks)**
4 Explain why the temperature is controlled in this investigation. **(2 marks)**

The blood

Blood is a tissue with red blood cells, white blood cells and platelets suspended in plasma.

Plasma

Blood **plasma** is a yellow liquid. It transports cells and platelets around the body through the **blood vessels**. It also transports dissolved substances, including:
- carbon dioxide from respiring cells
- soluble products of digestion, such as glucose and amino acids
- urea – a waste product formed when the liver breaks down excess proteins.

Centrifuging blood

The different parts in blood settle out quickly when blood is spun in a centrifuge.

- plasma (55%)
- white blood cells and platelets (<1%)
- red blood cells (45%)

Red blood cells

Most of the cells in blood are **red blood cells**.

These cells transport oxygen around the body through the blood vessels. They have several adaptations for this function.

Adaptation	Reason
contain **haemoglobin**	red pigment which binds reversibly to oxygen
no nucleus	provides more space to contain more haemoglobin
biconcave disc shape	increases the surface area to volume ratio for diffusion

biconcave (pushed in on both sides)

White blood cells

White blood cells are larger than red blood cells and have a **nucleus**. They are part of the body's **immune system**. There are different types of white blood cell, including:
- **phagocytes**, which engulf and destroy **pathogens** (disease-causing organisms)
- **lymphocytes**, which produce antibodies – proteins that bind to pathogens and lead to their destruction.

Platelets

Platelets are fragments of cells with no nucleus. They are involved in forming a blood clot at a cut or wound. Platelets are trapped when a meshwork of fibrin protein forms.

Worked example

The photo shows a thin smear of blood, viewed through a light microscope.
Identify A, B and C. **(3 marks)**

A is a platelet, B is a white blood cell and C is a red blood cell.

Platelets are much smaller than red cells or white cells. White cells have a nucleus (the dark area inside) but red cells do not. Blood contains more red cells than white cells.

A blood smear stained with a dye

Now try this

Compare the structure of red blood cells with the structure of white blood cells. **(4 marks)**

Blood vessels

The body contains three different types of **blood vessel**, called arteries, veins and capillaries.

Arteries

Arteries carry blood **A**way from the heart. The blood in arteries is under high pressure:
• This increases each time the heart beats.

Arteries have thick walls containing muscle and elastic fibres. These allow arteries to:
• stretch, without bursting, as blood is pushed through them
• regain their shape afterwards.

space where blood flows — thick wall of muscle and elastic fibres

Artery (cross-section)

Veins

Veins carry blood towards the heart. The blood in veins is under the lowest pressure, and is not pumped by the heart. Veins have:
• thinner walls than arteries
• a large **lumen** (space inside)
• **valves** to prevent blood flowing backwards.

Blood is squeezed through the veins by the contractions of surrounding skeletal muscle.

large space for blood to flow — thinner wall than artery

Vein (cross-section)

Capillaries

The **capillaries** are the smallest blood vessels, and contain blood at low pressure.

They form capillary beds – networks that supply blood to the body's tissues and organs.

Substances are exchanged between the blood in the capillaries and the cells around them.

Capillaries:
• are typically one blood cell wide
• have thin walls, one cell thick, allowing substances to diffuse easily through them.

→ waste products, e.g. carbon dioxide other cell products, e.g. hormones
→ substances needed by cells, e.g. oxygen, glucose

Worked example

The diagrams show part of a blood vessel in long section. The arrows indicate the direction of blood flow.
Identify the type of blood vessel, giving a reason for your answer.
(2 marks)

It shows a vein because there are valves, which are found in veins.

Blood can only flow one way between the flaps in the valve.

Now try this

1 List the three types of blood vessel, in order of decreasing blood pressure. **(1 mark)**

2 Explain why almost every body cell is very close to a capillary. **(3 marks)**

3 The inner wall of an artery is smooth. Suggest a reason for this adaptation. **(1 mark)**

The heart

The **heart** is an organ that pumps blood around the body.

Structure of the heart

Cells in the heart are supplied with oxygenated blood by the **coronary arteries**.

The diagram shows other structures in the heart.

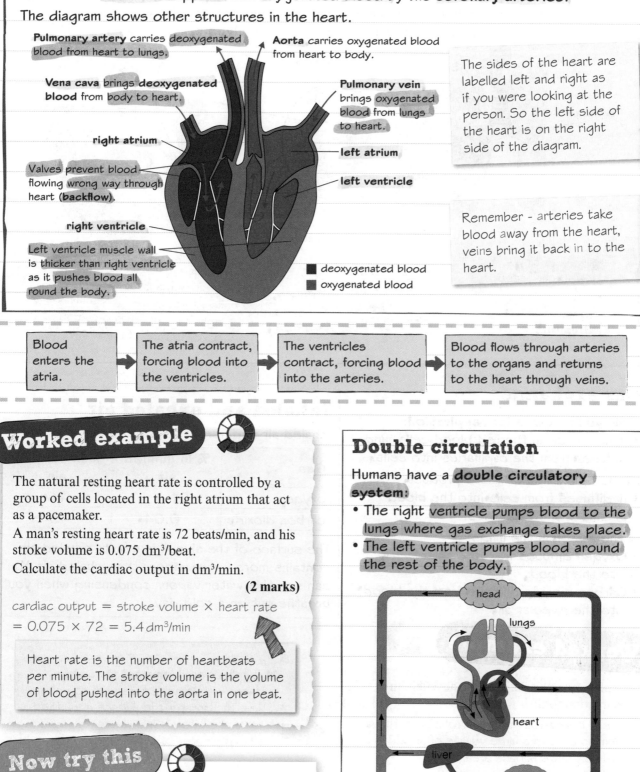

Pulmonary artery carries deoxygenated blood from heart to lungs.

Aorta carries oxygenated blood from heart to body.

Vena cava brings **deoxygenated blood** from body to heart.

Pulmonary vein brings oxygenated blood from lungs to heart.

right atrium

left atrium

left ventricle

Valves prevent blood flowing wrong way through heart (**backflow**).

right ventricle

Left ventricle muscle wall is thicker than right ventricle as it pushes blood all round the body.

■ deoxygenated blood
■ oxygenated blood

The sides of the heart are labelled left and right as if you were looking at the person. So the left side of the heart is on the right side of the diagram.

Remember - arteries take blood away from the heart, veins bring it back in to the heart.

Blood enters the atria. → The atria contract, forcing blood into the ventricles. → The ventricles contract, forcing blood into the arteries. → Blood flows through arteries to the organs and returns to the heart through veins.

Worked example

The natural resting heart rate is controlled by a group of cells located in the right atrium that act as a pacemaker.

A man's resting heart rate is 72 beats/min, and his stroke volume is 0.075 dm³/beat.

Calculate the cardiac output in dm³/min.

(2 marks)

cardiac output = stroke volume × heart rate

= 0.075 × 72 = 5.4 dm³/min

Heart rate is the number of heartbeats per minute. The stroke volume is the volume of blood pushed into the aorta in one beat.

Double circulation

Humans have a **double circulatory system**:

- The right ventricle pumps blood to the lungs where gas exchange takes place.
- The left ventricle pumps blood around the rest of the body.

head

lungs

heart

liver

gut

rest of body

Now try this

The muscular wall of the left ventricle of the heart is thicker than that of the right ventricle. Suggest an explanation for this difference. **(3 marks)**

The lungs

The lungs are part of the **respiratory system**, and are adapted for gaseous exchange.

Structure of the lungs

Air enters the mouth or nose and passes through the **trachea** (wind pipe):
- The trachea branches into two **bronchi**, one **bronchus** for each lung.
- These branch further and end in structures called the **alveoli**.

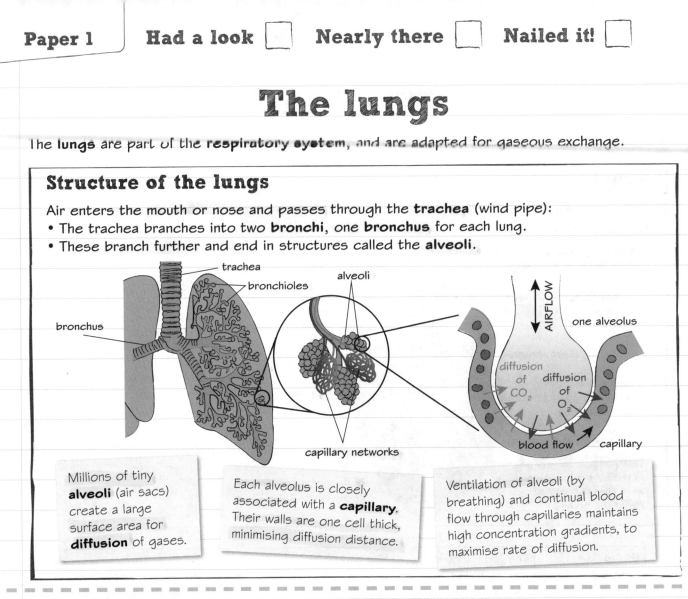

Millions of tiny **alveoli** (air sacs) create a large surface area for **diffusion** of gases.

Each alveolus is closely associated with a **capillary**. Their walls are one cell thick, minimising diffusion distance.

Ventilation of alveoli (by breathing) and continual blood flow through capillaries maintains high concentration gradients, to maximise rate of diffusion.

Gas exchange

Cells need oxygen for **respiration**:
- It is carried in the red blood cells, and diffuses from the capillaries into cells.

Carbon dioxide is a waste product:
- It diffuses from cells into the blood plasma in the capillaries.

In the alveoli in the lungs:
- Oxygen diffuses from the alveolar air into the blood.
- Carbon dioxide diffuses from the blood into the alveolar air.

Inhaled and exhaled air

Inhaled air is different to exhaled air:

Gas	% in inhaled air	% in exhaled air
oxygen	21	16
carbon dioxide	0.04	5

The surface of the alveoli is moist: exhaled air contains more water vapour than inhaled air. You can see this water vapour condensing when you breathe out on a cold day.

Worked example

Explain how the shape of the alveoli is an adaptation to their function. **(3 marks)**

Alveoli have a roughly spherical shape. This allows them to contain a large volume of air within a relatively large surface area.

Multicellular organisms need exchange surfaces with large surface area to volume ratios. The greater the surface area, the greater the rate of diffusion (alveoli give the lungs an area of 50–75 m²).

Describe **three** ways in which exhaled air differs in composition from inhaled air. **(3 marks)**

Cardiovascular disease

Cardiovascular disease is a general term for conditions affecting the heart or blood vessels.

Coronary heart disease

Fatty material can build up inside the coronary arteries, the arteries that supply blood to the heart itself. This reduces:

- blood flow in the arteries
- the amount of oxygen for the heart muscle
- it also causes pain, heart attacks or death.

A narrowed artery can be treated using a stent. This is a wire frame inserted into the damaged artery. It is expanded using a tiny balloon, which is then removed allowing blood to flow more freely.

— narrowed blood vessel

— stent in place inside vessel

Types of treatment

Cardiovascular diseases are non-communicable diseases. They are not caused by an infectious pathogen, so they cannot be passed from person to person.

Cardiovascular diseases usually get slowly worse over a long time. Treatments include:

- drugs
- artificial devices
- transplants.

An artificial pacemaker is a device implanted under the skin. It sends electrical impulses through a wire to the heart. This corrects irregularities in heart rate.

Faulty heart valves

The heart valves stop blood flowing in the wrong direction through the heart. If they become faulty, the heart becomes less efficient at pumping blood to the lungs and the rest of the body. Faulty valves may:

- not open fully
- develop a leak.

This causes tiredness and breathlessness, and can lead to death.

Faulty heart valves can be replaced by:

- mechanical valves made from metals or polymers
- biological valves, from human donors, cows or pigs.

Heart failure

Heart failure is when the heart can no longer pump enough blood for the body's needs. Treatment can include:

- heart transplant
- heart and lungs transplant.

Artificial hearts are sometimes used to keep patients alive until their heart transplant, or to rest the heart to help recovery. Treatments like this are expensive, with significant risks.

Patients have to take drugs to prevent blood clots caused by mechanical valves and hearts, or to prevent rejection of biological valves by the immune system.

Worked example

Statins are drugs that reduce blood cholesterol levels. Explain why this decreases the risk of developing heart disease or suffering a heart attack.
(2 marks)

If the cholesterol level is reduced, the rate at which fatty material builds up in the arteries is also reduced. This means that the coronary arteries are less likely to become blocked.

Statins are relatively cheap drugs. However, they can produce side effects, such as muscle pain and an increased risk of diabetes.

Statins and other drugs have drawbacks as well as benefits. The results of research suggests that the benefits greatly outweigh the risks of harm.

Now try this

Suggest why a person with serious heart disease may only be offered surgery if they have not responded to other treatments.
(2 marks)

25

Health and disease

Diseases are major causes of ill health.

What is health?

Health is the state of physical and mental wellbeing.

health — mental wellbeing

such as how you feel about yourself

such as eating and sleeping well, and being free from disease

physical wellbeing — social wellbeing

such as how well you get on with other people

Types of disease

 Non-communicable diseases.
These are not passed from person to person.

2 **Communicable** diseases.
These can be passed from one person to another. You can revise pathogens and the diseases they cause on pages 34–36.

Non-communicable	Communicable
Number of cases changes gradually over time.	rapid variation in the number of cases over time
Cases may be widespread.	Cases are often localised.
e.g. cardiovascular disease, cancer, diabetes	e.g. colds, flu, measles, food poisoning

Factors affecting health

Diseases are not the only factors that affect health. These factors also affect health:
- diet
- stress
- life situations.

Diseases may also interact with one other.

Disease or problem	Additional problem
defects in the immune system	more prone to infectious diseases
immune reaction caused by a pathogen	can trigger allergies, e.g. skin rashes, asthma
severe physical ill-health	can lead to mental illness such as depression
viruses in cells	can be a trigger for cancer

 Worked example

A course of antibiotics kills *H. pylori* and most ulcers heal in 1–2 months. If ulcers are caused by aspirin, doctors can prescribe other painkillers.

Helicobacter pylori is a bacterium that lives in the stomach lining. People usually become infected during childhood, probably from family members. About 20% of people in the UK have *H. pylori*. Of these, about 15% develop stomach ulcers. Long-term use of high doses of aspirin also cause stomach ulcers. Evaluate whether stomach ulcers are a communicable disease.

(5 marks)

H. pylori is caught from other people. If most people with ulcers also have this bacterium, stomach ulcers would be a communicable disease. On the other hand, most infected people do not develop ulcers, and people can also develop ulcers because they take a lot of aspirin over a long time. It appears that stomach ulcers might be a communicable disease, but not enough information is given to be sure.

Now try this

A stroke can happen when a blood clot blocks an artery in the brain and nerve cells die due to lack of oxygen. Explain whether a stroke is a communicable or a non-communicable disease. **(2 marks)**

Lifestyle and disease

Lifestyle factors such as diet are linked to increased rates of disease.

Exercise and disease

Obesity (being very overweight) is a lifestyle risk factor for type 2 **diabetes** (you can revise diabetes on page 63). A person who exercises regularly is more likely to control their body mass. They are also less likely to suffer from **cardiovascular disease** (see page 25).

benefits of exercise

better weight control → better health

Inherited factors

Inherited factors are not lifestyle factors, but they can affect health. For example, some people inherit certain **alleles** that increase their blood cholesterol level. They may have higher cholesterol levels than people without these alleles and who eat the same diet.

When we study the effects of lifestyle factors on disease, we must compare people with similar inherited factors to obtain valid results.

Some other risk factors

Many diseases are caused by the interaction of different factors. Some risk factors have a proven **causal mechanism**. Scientists:

* can show that the risk factor is a cause of the disease, and
* can show how this works.

For example, risk factors with proven causal mechanisms for cancers include:

* toxic substances called **carcinogens**
* exposure to ionising radiation.

Scientists have also identified lifestyle risk factors and genetic risk factors for cancers.

Cancer

Cancer is caused by changes in cells that lead to uncontrolled cell growth and division.

1   **Benign tumours** are growths of abnormal cells. These cells are contained in one area, usually inside a membrane. They do not invade other parts of the body.

2   **Malignant tumours** are cancers. They invade neighbouring tissues. They also spread to different parts of the body in the blood, forming **secondary tumours**.

Worked example

In a study, 522 overweight people of similar age who were at risk of developing diabetes were separated into two groups.

Group A was given advice on how to live more healthily. Group B, the control group, received no such advice. After four years, 11% of Group A and 23% of Group B had developed diabetes.

It was concluded that living healthily reduces the risk of developing diabetes. Explain the extent to which the results support this conclusion.

(3 marks)

The results support the conclusion. Only 11% of the group who received advice developed diabetes, just under half of the percentage in the control group. The results are likely to be valid because the study used many people, of similar age, and who all started with the same problems.

> The validity of a study increases as the number of people involved increases, and by having a control group.

Now try this

A study into the effects of physical fitness found that unfit overweight men had a much higher risk of type 2 diabetes than fit overweight men.
Suggest another suitable group of men for this study.

(1 mark)

Alcohol and smoking

Drinking **alcohol** and **smoking** are lifestyle risk factors for some non-communicable diseases.

Effects of alcohol

Ethanol is the alcohol found in alcoholic drinks.

> Drinking a lot of alcohol at one time can cause **short-term harm** including:
> - blurred vision
> - slow reactions
> - lowered inhibitions so you take more risks.

> Drinking a lot of alcohol over a long period can cause **long-term harm** including:
> - brain damage
> - **liver cirrhosis.**

Drinking alcohol while pregnant can lead to long-term harm to the baby, including:

👎 poor growth

👎 abnormal facial features

👎 behaviour and learning problems.

The risk increases as consumption increases.

The effects of smoking

Tobacco smoke contains many different harmful substances.

Nicotine is addictive.

This can lead to low birth weight in babies whose mothers smoke.

Carbon monoxide reduces how much oxygen the blood can carry.

Substances in cigarettes can cause blood vessels to narrow, increasing blood pressure. This can lead to **cardiovascular diseases** such as heart attack or strokes.

Chemicals in **tar** are **carcinogens** that cause cancers, particularly of mouth and lungs.

Worked example

The graph shows the incidence of liver disease and the mean ethanol consumption per person per year for different countries. Describe the relationship and give a reason that explains it. **(2 marks).**

As the amount of alcohol consumed increases, the incidence of liver disease also increases. This is because ethanol is poisonous, particularly to liver cells.

This graph shows a **positive correlation** between two factors. You might also see a graph where one factor increases while the other decreases – a **negative correlation**.

Now try this

1 Give **two** effects of long-term excessive alcohol consumption. **(2 marks)**

2 Explain why smoking increases the risk of having a heart attack. **(2 marks)**

The leaf

The leaf is a plant **organ** that contains different tissues.

Plant organs

Plant organs include:
- the **leaf**, which carries out photosynthesis
- the **stem**, which supports leaves and flowers
- the **root**, which anchors the plant in the soil, and absorbs water and mineral ions from the soil water.

These organs form a plant **organ system** for transport of substances around the plant.

epidermis
mesophyll
xylem
phloem

Leaf tissues

cuticle (thin waxy layer)
chloroplast
xylem (transports water)
phloem (transports dissolved sugar)
air space
cuticle
guard cell
stomata
upper epidermis
palisade mesophyll
mesophyll tissue
spongy mesophyll
lower epidermis

Worked example

Describe how the mesophyll tissue is adapted for photosynthesis.   **(4 marks)**

The palisade mesophyll cells contain many chloroplasts. Their cylindrical shape means that light has few cell walls to pass through to reach the chloroplasts.

The spongy mesophyll cells pack together in an irregular way so that there are air spaces, with large surface areas for gas exchange.

The mesophyll is sandwiched between the upper and lower epidermis. These tissues consist of thin, flat cells that protect the leaf but easily let light through.

Spongy mesophyll cells have fewer chloroplasts than palisade mesophylls. They are found on the lower surface of the leaf, which receives less sunlight.

Now try this

Only the guard cells in the epidermis tissues can carry out photosynthesis. Use the diagram of the leaf tissues to give a reason why.
(2 marks)

Transpiration

Transpiration is the movement of water from roots to leaves, where it evaporates into the air.

Transpiration stream

The **transpiration stream** is the continuous movement of water through a plant. It is driven by the **evaporation** of water from the leaves. The transpiration stream also carries dissolved **mineral ions.** You can revise adaptations of root hair cells and xylem tissue on page 5.

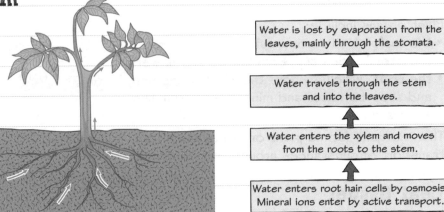

Water is lost by evaporation from the leaves, mainly through the stomata.

Water travels through the stem and into the leaves.

Water enters the xylem and moves from the roots to the stem.

Water enters root hair cells by osmosis. Mineral ions enter by active transport.

Stomata

The **stomata** are openings on the surface of a leaf. They are found mainly on the lower surface:
- Most evaporation from a leaf happens through its stomata.
- The size of the opening in a **stoma** is controlled by its **guard cells**.
- The stomata control gas exchange and water loss.

When guard cells take in water by osmosis, they become turgid (swollen) which causes the stoma to open.

When guard cells lose water, they become flaccid and the stoma closes.

chloroplast — vacuole
cell wall —
stoma —
— nucleus

stoma open   stoma closed

Worked example

Four leaves were cut from a plant. Their cut ends were sealed with grease. The leaves were covered in grease in the following ways.
- A – Neither surface was greased.
- B – The upper surface was greased.
- C – The lower surface was greased.
- D – Both surfaces were greased.

The mass of each leaf was measured. They were left for the same time before measuring their mass again. The table shows the results.

Leaf	% loss in mass
A	40
B	37
C	4
D	2

Explain these results.   **(4 marks)**

The loss in mass is due to evaporation of water. Leaves A and B lost the most water vapour because it could still escape through the stomata on the lower surface. Leaves C and D lost very little water because their stomata were covered.

Environmental factors also affect the rate of transpiration (see page 31).

Environmental factors also affect the rate of transpiration (see page 31).

Now try this

1 Describe how water moves into a plant and escapes from the leaf.   **(4 marks)**

2 Guard cells become turgid in the daytime when it is light, and flaccid when it is dark at night. Explain the effect of this change on the rate of transpiration.   **(4 marks)**

Investigating transpiration

Environmental factors affect the rate of transpiration.
You can measure the **transpiration rate** using a <u>**potometer**</u>.

1 Record the starting position of the bubble in the capillary tube using the scale.

2 Record the position of the bubble after allowing the plant shoot to transpire for a known time.

3 Calculate the rate at which the bubble moved. This gives the rate of water uptake, which gives the rate of water loss and so the rate of transpiration.

reservoir for pushing air bubble to right-hand end of capillary tube

rubber stopper

capillary tube with scale

air bubble

Factors affecting transpiration rate

Factor	Effect on transpiration rate if factor increases	Explanation of effect
light intensity	increases	• Rate of photosynthesis increases, so the stomata are more open for gas exchange.
temperature	increases	• Rate of photosynthesis increases, so the stomata are more open for gas exchange. • Water molecules move quickly, so rate of diffusion increases.
air movement	increases	• Concentration gradient of water vapour increases, so rate of diffusion increases.

Worked example

A plant shoot was placed in a potometer. The position of the bubble was recorded before, and after, leaving the shoot in still air for 5 minutes. The potometer was reset, and the experiment repeated with the shoot enclosed in a plastic bag. The table shows the results.

	Outside bag	Inside bag
Start reading (mm)	2	5
End reading (mm)	52	25
Distance (mm)	50	20
Rate (mm/min)	50/5 = 10	20/5 = 4

(a) Complete the table. **(2 marks)**

(b) Explain the results. **(2 marks)**

The humidity inside the bag increases as water vapour escapes from the leaves. As the humidity increases, the concentration gradient of water vapour decreases. This reduces the rate of diffusion, so the transpiration rate decreases.

Maths skills You can calculate the **volume** of water taken up by the shoot, in mm³, if you know the diameter of the capillary tubing in mm:

volume = distance × πr^2

where radius, r = diameter/2

Now try this

1 State four environmental factors that affect the rate of transpiration. **(4 marks)**

2 A plant wilts and becomes floppy if it loses water faster than the roots can absorb water. Explain why a plant is more likely to wilt when the weather is hot and windy, rather than cool and still. **(5 marks)**

Translocation

Translocation is the movement of dissolved food molecules through phloem tissue.

Transport through phloem

Phloem tissue transports dissolved sugars from the leaves to the rest of the plant.

These sugars are for immediate use, for example in respiration and for storage.

Phloem tissue consists of tubes of elongated cells (see page 5). Cell sap can move from one phloem cell to the next through pores in the end walls.

dissolved sucrose needed for growth in growing regions, e.g. bud

Sucrose is produced in leaves from glucose formed during photosynthesis.

phloem

Dissolved sucrose is carried around the plant in phloem.

storage organ (e.g. potato)

dissolved sugars converted to starch and stored in storage organs so they can be used later

Worked example

Carbon dioxide containing radioactive carbon atoms is supplied to a plant's leaf. The plant is left in the light for several hours. Its stem is then analysed to detect radioactivity.

Explain why radioactivity is detected in the stem.

(2 marks)

Carbon dioxide is used by the leaf for photosynthesis. This means that the glucose formed contains radioactive carbon atoms. Some glucose is converted to sucrose. This is transported through the plant in the phloem, so radioactivity is detected in the stem.

Some glucose is used in respiration. Some is converted into storage substances such as starch. Radioactive starch would be detected in this plant.

Meristem tissue

Meristem tissue is found at the growing tips of shoots and roots. It consists of **stem cells**.

These cells can **differentiate** to form specialised cells, including those found in xylem and phloem tissue.

You can revise stem cells on page 10.

leaf

shoot tip

area where cells are dividing rapidly to make more new cells

zone of **elongation**

differentiation of cells to form xylem and phloem

Now try this

Transport can take place in both directions in phloem tissue. Explain why this is useful to a plant during the night when it is dark.

(4 marks)

Extended response – Organisation

There will be at least one 6-mark question on your exam paper. For these questions, you will need to think scientifically and structure your answer logically, showing how the points you make are related to each other.

For the questions on this page, you can revise the circulatory system and the lungs on pages 21–24.

Worked example

The circulatory system transports substances to and from the tissues in the body.
Explain how this organ system is adapted to supply oxygen and to remove carbon dioxide.

The blood contains red blood cells suspended in plasma. Haemoglobin in the red blood cells binds to oxygen so it can be transported around the body. Carbon dioxide dissolves in the plasma so it can be transported around the body.

The heart has muscles which can contract in an organised way to pump blood. Deoxygenated blood enters the right side of the heart through the vena cava. It is pumped through the pulmonary artery to the lungs. In the lungs, oxygen diffuses from the air into the blood and carbon dioxide diffuses from the blood into the air.

Oxygenated blood returns to the left side of the heart through the pulmonary vein. The heart contains valves so the blood flows in one direction only. The heart pumps blood through the aorta to the rest of the body.

The various arteries carry oxygenated blood to the capillaries in body tissues. The capillaries have thin walls and a large surface area to allow efficient gas exchange between the blood and cells in the tissues.

Command word: Explain

When you are asked to **explain** something, you must include reasoning or justification of the points you make.

Blood also contains white blood cells and platelets, but these have other functions and need not be mentioned.

The answer could also mention that a group of cells located in the right atrium act as a pacemaker. They control the natural resting heart rate.

The answer could also state that this organ system is a double circulatory system. This means it can maintain a high blood pressure and flow rate to the rest of the body.

The answer could also explain that blood leaves the capillaries and enters the veins, so it can return to the heart.

You need to show comprehensive knowledge and understanding, using relevant scientific ideas, to support your answer. Make sure your answer links different points in a logical way.

Now try this

Include factors that affect the rate of diffusion (you can revise these on page 11). Explain how the structure of alveoli is related to each of these factors.

The diagram shows alveoli, structures found in the lungs. Explain how the alveoli are adapted for efficient gas exchange between air in the lungs and blood in the capillaries. **(6 marks)**

oxygenated blood out

air in and out

deoxygenated blood in

Viral diseases

Pathogens are microorganisms that cause infectious disease. **Viruses** are a type of pathogen.

Viruses

A virus consists of genetic material (DNA or RNA) surrounded by a coat of proteins and lipids.

Viruses:
- are much smaller than cells
- exist and reproduce inside cells, causing cell damage
- reproduce rapidly in the body
- can infect plants or animals
- are spread by direct contact, water or air.

Colds and flu are viral diseases.

Virus life cycle

Virus enters the cell through the cell membrane.
↓
Cell produces more viral DNA or RNA, and viral proteins.
↓
New viruses are assembled inside the cell.
↓
New viruses leave or burst out of the cell.

Human viral diseases

Eg. Tobacco mosaic virus

The table shows some information about two human viral diseases.

	Measles virus	HIV — Human immunodeficiency virus
How it is spread	• inhalation of droplets from sneezes and coughs	• sexual contact • exchange of body fluids such as blood, e.g. when drug users share needles
Symptoms of infection	• fever and a red skin rash • can be fatal if complications occur (so most young children are vaccinated against measles)	• flu-like illness at the start • attacks body's immune cells unless it is controlled with antiretroviral drugs • AIDS (late-stage HIV infection) occurs when the body's immune system is too badly damaged to deal with other infections or cancers.

Worked example

Tobacco mosaic virus, TMV, is a plant pathogen. Describe the symptoms of infection by TMV.   **(2 marks)**

The leaves get a distinctive mottled or 'mosaic' pattern of discoloration. This reduces the rate of photosynthesis, so the growth of the plant is affected.

TMV affects tomato plants and many other species of plants, not just tobacco plants.

The virus can spread by contact from cell to cell, and plant to plant. It can also travel through the phloem.

Now try this

1 Give **two** ways in which viruses can be spread.   **(2 marks)**

The host cell is the cell the virus infects.

2 Explain how viruses damage their host cells.   **(2 marks)**

Bacterial diseases

Bacteria are **prokaryotic** organisms that can act as pathogens.

Bacteria

A bacterial cell contains cytoplasm, a loop of DNA and one or more **plasmids** (rings of DNA) with a cell membrane surrounded by a cell wall.

Bacteria:
- are usually smaller than plant and animal cells, but larger than viruses
- may reproduce rapidly inside the body
- may produce **toxins** – poisons that damage tissues, making us feel ill
- can infect plants or animals
- are spread by direct contact, water or air.

Cholera and tuberculosis (TB) are bacterial diseases.

Ignaz Semmelweis (1818–1865)

Semmelweis wondered why many women died of infection soon after childbirth.

> Semmelweis realised that doctors might be transferring infection between patients on their hands.

> He insisted that doctors wash their hands before examining each patient.

> Death rates fell rapidly in the wards where doctors washed their hands.

Pathogens were unknown then, so he could not explain why hand washing worked.

Human bacterial diseases

Eg.

The table shows some information about two human bacterial diseases.

	Salmonella food poisoning	Gonorrhoea
How it is spread	• food containing bacteria • bacteria on food prepared in unhygienic conditions	• sexual contact
Symptoms of infection	• fever • abdominal cramps • vomiting • diarrhoea	• thick yellow or green discharge from the vagina or penis • pain on urinating

Gonorrhoea is a sexually transmitted disease (STD).

(a) Explain why its spread can be controlled using antibiotics. **(2 marks)**

Gonorrhoea is a bacterial disease and antibiotics can kill infective bacteria inside the body.

(b) Give one other way in which the spread of gonorrhoea may be controlled. **(1 mark)**

People could use condoms during sexual intercourse.

Gonorrhoea was easily treated in the past using penicillin, an antibiotic. However, many resistant strains of bacteria appeared. You can revise antibiotic resistance on page 93.

Around 1 in 10 infected men and almost half of infected women do not have any symptoms. So people may spread the pathogen without knowing they are infected.

1 Give **one** way in which bacterial toxins cause harm. **(1 mark)**

2 Explain why you should wash your hands before handling food. **(2 marks)**

Fungal and protist diseases

Fungi and protists are **eukaryotic** organisms that may act as pathogens.

Fungi   Eg. Rose black spot

Single-celled fungi include yeasts and moulds. They have cell walls but no chloroplasts. Some may produce toxins that make people ill.

> You can revise the use of yeast cells in fermentation to make bread and alcoholic drinks on page 49.

Protists

Protists do not form tissues. They exist as a huge variety of unicellular organisms.

Some protist species are **parasites**: they benefit at the expense of the host they infect.

Malaria

The symptoms of **malaria** include recurring episodes of fever and death.

The pathogens that cause malaria are protists.

They are parasites with a **life cycle** that involves an insect called a **mosquito**.

The mosquito is a **vector** – it carries the pathogen from person to person.

The spread of malaria is controlled by:
- preventing mosquitos breeding
- using mosquito nets at night to avoid being bitten.

infected liver cells → infected red blood cells

Infected mosquito bites a human and injects the protist that causes malaria.

Mosquito bites a second person and spreads malaria.

Mosquito becomes infected with parasite when it takes a blood meal.

Worked example

Rose black spot is a fungal disease. It causes purple or black spots to develop on leaves. These often turn yellow and drop early.

(a) Explain why rose black spot affects the growth of infected plants. **(2 marks)**

Photosynthesis is reduced because of the damage to the leaves, so the plant cannot make enough food for healthy growth.

(b) Give one way in which rose black spot may be treated. **(1 mark)**

Spray a fungicide onto the plants.

> Rose black spot is spread in the environment by fungal spores that travel in the wind, or by water in rain splashes.

> Fungicides are substances that kill fungi. They are used in some treatments for fungal diseases that affect people, such as athlete's foot.
>
> Another way to treat rose black spot is to remove and destroy the affected leaves.

Now try this

1 Ringworm is a fungal disease spread through contact with infected skin, or with contaminated clothes. Suggest **two** ways to prevent its spread. **(2 marks)**

2 A common drug to treat malaria kills malaria parasites in red blood cells.
   (a) Explain how this drug may prevent infection with the malaria parasite. **(2 marks)**
   (b) Explain how this drug may reduce the spread of malaria. **(2 marks)**
   (c) Mosquitos breed in still water. Explain why authorities may drain ditches in areas affected by malaria. **(2 marks)**

Human defence systems

The human body has non-specific defence systems against **pathogens**. These systems are found in various places including the nose, trachea and bronchi, stomach and skin.

Chemical and physical defences

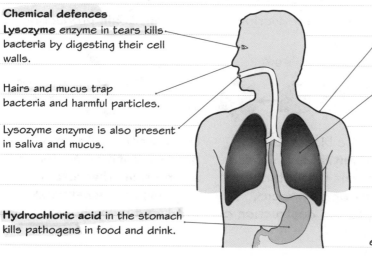

Chemical defences

Lysozyme enzyme in tears kills bacteria by digesting their cell walls.

Hairs and mucus trap bacteria and harmful particles.

Lysozyme enzyme is also present in saliva and mucus.

Hydrochloric acid in the stomach kills pathogens in food and drink.

Physical barriers

Unbroken **skin** forms a protective barrier because it is too thick for most pathogens to get through.

Sticky **mucus** in the trachea and bronchi traps pathogens. **Cilia** on the cells lining these passages move in a wave-like motion, moving mucus and trapped pathogens out of the lungs towards the back of the throat where it is swallowed.

Mucus traps pathogens.

Cilia move mucus away from lungs.

cilia

epithelial cells

Chemical defences involve substances that kill pathogens and make them inactive.

Physical defences make it more difficult for pathogens to enter the body.

> Remember that epithelial cells line the surface of tubes.

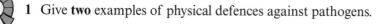

Worked example

Substances in cigarette smoke paralyse the cilia in the epithelium of the respiratory system. Explain why smokers are more likely to suffer from lung infections than non-smokers. **(3 marks)**

Mucus in the trachea and bronchi traps bacteria. The cilia move this mucus up into the throat, where it is swallowed and destroyed. When the cilia are paralysed, this no longer happens, so bacteria and other pathogens are no longer removed.

> Smokers develop a 'smoker's cough'. Their lungs become clogged with mucus which they have to cough up.

Skin defences

The skin is the organ that covers the outer surface of the body. It acts as a barrier to pathogens and has other features that defend against them:

1 It secretes antimicrobial substances that kill pathogenic bacteria or inhibit their growth.

2 Many species of non-pathogenic bacteria live on the skin. They secrete substances that kill pathogenic bacteria, and compete with them for nutrients.

3 Scabs form over damaged skin, keeping pathogens out while the skin repairs itself.

Now try this

1 Give **two** examples of physical defences against pathogens. **(2 marks)**

2 Explain why food poisoning might be more likely when a person swallows their food without chewing it thoroughly first. **(3 marks)**

> What defences are present in the mouth and stomach?

The immune system

It a pathogen enters the body, the immune system tries to destroy the pathogen.

White blood cells

White blood cells help to defend against **pathogens** in different ways.

lymphocytes

phagocytosis

white blood cell

pathogen

Some white blood cells (engulf)* **ingest** pathogens. They flow around them and destroy them.

antibody production

antibodies

pathogen

Some white blood cells produce **antibodies**. These are proteins that attach to **antigens** on pathogens, leading to the destruction of the pathogen.

antitoxin production

toxins from pathogen

antitoxins

Some white blood cells produce **antitoxins**. These are proteins that attach to poisonous substances produced by pathogens, making them **inactive**.

Antibodies

Antibodies recognise and attach to antigens (substances produced by organisms):

1 A white blood cell with antibodies specific to an antigen attaches to that antigen.

2 The white blood cell is activated, and divides many times to produce **clones**. These cells produce lots of antibodies against the antigen.

3 Some of the white blood cells remain in the blood, as 'memory lymphocytes', when the infection is over.

Immunity

White blood cells respond to infection by making antibodies.

Amount of antibody in bloodstream in arbitrary units

first infection

second infection

immunity level

Time in weeks

The memory lymphocytes are made after the infection and then respond more quickly to another infection by the pathogen.

An antibody is **specific** for a particular antigen.

Measles is a viral disease. Explain why you do not get measles more than once. **(2 marks)**

After the first infection with the measles virus, memory cells remain in the blood. If the virus infects the body again, these white blood cells cause a faster and larger immune response so you do not fall ill.

The immune response to the first infection is called the **primary response**. The immune response to a later infection is called the **secondary response**.

Describe **three** ways in which white blood cells defend the body against pathogens. **(6 marks)**

Vaccination

Vaccination prevents illness in individuals and reduces the spread of pathogens.

Vaccines

Different **vaccines** give protection against bacterial disease such as diphtheria and tetanus, or against viral diseases such as measles, mumps and polio.

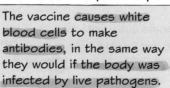

| A vaccine contains a small amount of a dead or inactive form of a pathogen. | → | The vaccine causes white blood cells to make antibodies, in the same way they would if the body was infected by live pathogens. | → | If the live pathogen infects you later, your immune system remembers how to destroy it, and responds quickly so you don't fall ill. You are immune. | memory lymphocytes! |

Depending on the particular vaccine, vaccination involves swallowing the vaccine, or having an injection into the skin or muscle.

Benefits and risks of vaccination

👍 Vaccinated people become immune to a disease without getting it.

👍 The spread of pathogens is reduced when a large proportion of the population is immunised (**herd immunity**).

👎 Some people develop side effects such as soreness and swelling, or show mild symptoms of the disease.

👎 Some vaccinations, particularly against flu, only give partial protection if there are different strains of the pathogen.

Worked example

The MMR vaccine protects against measles, mumps and rubella (German measles). Children are given the vaccine when they are one year old, and again before starting school in case they did not become immune to measles the first time. The graph shows how the incidence of measles and the percentage of vaccinated children changed between 1996 and 2008.

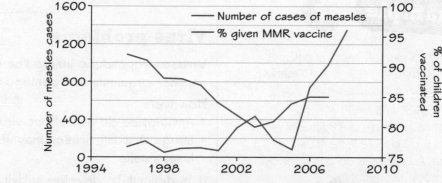

Use the graph to explain the importance of vaccination in preventing disease. **(3 marks)**

The percentage of children given the MMR vaccine decreased from over 90% in 1996 to about 80% in 2003. Although the percentage began to increase again, the number of measles cases increased rapidly after 2005. This suggests that the risk of being infected increases when the proportion of children vaccinated decreases.

Now try this

1 Describe the difference between a vaccine and vaccination. **(2 marks)**

2 If many people have been vaccinated to stop them catching an infectious disease, people who have not been vaccinated are less likely to catch it. Explain why this is. **(3 marks)**

Antibiotics and painkillers

Painkillers are medicines that treat **symptoms** of a disease but do not kill pathogens.

Antibiotics

An **antibiotic** is a medicine that kills infective bacteria inside the body:

Symptoms are the results of a disease, such as aches and pains, and having a high temperature.

- This helps to cure bacterial diseases.
- Specific bacteria are only killed by specific antibiotics, so doctors must prescribe the correct antibiotic.
- 👍 The use of antibiotics has greatly reduced deaths from infectious bacterial diseases.
- 👎 Strains of bacteria that are resistant to antibiotics have emerged.

Penicillin is an example of an antibiotic (you can revise the discovery of medicines on page 41).

Antibiotic resistance

Some strains of bacteria have developed a resistance to antibiotics. Certain antibiotics no longer kill these bacteria. MRSA is a strain of bacteria that has developed resistance to several antibiotics.

| **Mutation** can produce new **antibiotic-resistant** strains of bacteria. | When the antibiotic is used, non-resistant bacteria die, but resistant bacteria survive and reproduce. | The population of resistant bacteria increases. Infections can only be treated with a new antibiotic. | If there is no new antibiotic to control the infection, it may spread rapidly, causing an epidemic or pandemic. |

This is an example of natural selection.

You can revise natural selection on pages 84 and 85.

Worked example

Explain how antibiotics can kill bacteria without damaging the body's tissues. **(3 marks)**

Antibiotics can prevent processes that happen in bacterial cells, but not in animal cells. For example, some antibiotics stop the bacteria forming their cell walls. This prevents them growing or dividing. Animal cells do not have cell walls, so they are not affected.

Some antibiotics are toxic to some animals, e.g. penicillin is toxic to guinea pigs.

Virus problems

Viruses reproduce inside the cells of another organism and damage these cells.

However:
- Antibiotics do not affect viruses.
- Drugs that kill viruses may also harm human cells.

It is difficult to develop antiviral drugs that do not also damage the body's tissues, so viral diseases can be difficult to treat.

Now try this

1 Explain why antibiotics are not effective against measles. **(2 marks)**

What sort of pathogen causes measles?

2 Explain why antibiotics are no longer used to treat non-serious bacterial infections. **(3 marks)**

New medicines

New medical drugs have to be tested and trialled to check that they are safe and effective.

Discovering drugs

A medical **drug** is a substance that treats the cause of a disease, or its symptoms.

Drugs were traditionally extracted from plants and microorganisms (see box on the right).

Most new drugs are **synthesised** (made using chemical reactions) by scientists working in the **pharmaceutical industry**.

Traditional sources of drugs

Drug	Origin	Treats
digitalis (digoxin)	foxglove plants	irregular heartbeat
aspirin	willow bark and leaves	aches and pains
penicillin	*Penicillium* mould	bacterial diseases

discovered by Alexander Fleming (1881–1955)

Drug testing

New drugs are tested extensively to determine their:
- **toxicity** – how poisonous or harmful they might be
- **efficacy** – how well they work at treating a disease
- **dose** – how much drug must be given for it to work.

1 Preclinical testing

New drugs are tested in the laboratory using cells and tissues. Promising substances are then tested on live animals.

cultures of cells   cultures of tissues   animals

2 Clinical trials

The next stages use people:
- Very low doses are given to healthy volunteers to check that the drug is <u>not toxic</u> to people.
- If the drug is found to be safe, different doses are given to patients to test its efficacy and **optimum dose** (the dose that works the best).

Worked example

Double blind trials may be used in clinical trials with patients. Explain what is meant by a double blind trial. **(3 marks)**

Some patients are given the drug and some are given a placebo. This is designed to look just like the drug, but it does not contain any of the drug. The doctors and patients do not know who is given the drug and who is given the placebo.

~ active ingredient

It is possible for people to feel better after taking a medicine, just because they expect to feel better. This is the **placebo effect.**

Double blind trials try to minimise this effect. The doctors and patients only find out who received the placebo after the trial. This way, differences between the two groups of patients should be due to the drug.

Now try this

1 Describe the origin and use of one drug traditionally extracted from plants. **(2 marks)**

2 In a blind trial, the patients do not know who receives the placebo, but the doctors do. Suggest a reason why this type of trial is less satisfactory than a double blind trial. **(2 marks)**

Monoclonal antibodies

Monoclonal antibodies are identical antibodies produced by cloned cells.

Making monoclonal antibodies

Monoclonal antibodies are produced by **hybridoma** cells, produced by fusing a lymphocyte cell with a tumour cell.

Inject the mouse with an antigen

lymphocytes (white blood cells) taken from mice

lymphocytes stimulated to make a particular antibody

tumour cell (can divide but not produce antibodies)

lymphocyte cells fused (combined) with tumour cell

hybridoma cell

Hybridoma cell can divide and produce antibodies. (mitosis)

A single hybridoma cell can be **cloned**:
- Many identical cells are produced.
- All produce the same antibody, which is specific for one protein **antigen**.
- Large amounts of the monoclonal antibody can be collected and purified.

Uses of monoclonal antibodies

Monoclonal antibodies can be used in several ways:

1. **Diagnosis**, such as in pregnancy tests. The monoclonal antibodies attach to a pregnancy hormone present in urine, causing a reaction with an indicator.

2. In laboratories, to measure the levels of **hormones** and other substances in blood, or to detect **pathogens**.

3. In scientific research, to locate or identify specific molecules in a cell or tissue. Monoclonal antibodies, carrying a fluorescent dye, bind to these molecules.

dye molecule

monoclonal antibody molecule

This part binds to an antigen molecule.

Now try this

Describe how monoclonal antibodies could be used to deliver a radioactive substance to a particular type of cell.   **(4 marks)**

Radioactive substances can be chemically joined to proteins, including antibody molecules.

Plant disease

Plants can be infected by a range of viral, fungal and bacterial pathogens.

Detecting plant diseases

Farmers and gardeners detect diseases in their plants in several ways.

change in the normal appearance of the plant e.g. discoloured flowers or spots on leaves

presence of pests infesting the plant, e.g. aphids   discolor leaf

Identify the visible symptoms

death of parts of the plant, e.g. dead leaves or roots

poor development of part of the plant, e.g. small leaves

Different diseases may give the same symptoms, so other tests are needed to identify the cause.

Insect **pests** such as aphids are usually easily seen on stems and leaves.

Examples of diseases

Disease	Pathogen
tobacco mosaic virus, TMV	virus
rose black spot → black spot	fungus
bacterial wilt	bacteria

You can revise the symptoms of TMV infection on page 34, and the symptoms of rose black spot infection on page 36.

Aphids are insect pests that damage plants. Their sharp mouthparts can reach phloem, so they can feed on the sugary solution there. Aphids can also act as disease **vectors** if their mouthparts carry plant pathogens.

Identifying diseases

A plant disease can be identified in several ways, including:

1. referring to a gardening manual or gardening website to compare the detailed symptoms

2. taking infected plants to a laboratory for testing, such as microscopic examination and bacterial culture

3. using testing kits which contain monoclonal antibodies that are specific for a particular pathogen.

Worked example

Plants can be damaged by a range of mineral ion deficiency conditions.

(a) Give two ions important for plant growth.   **(2 marks)**

Nitrate ions, NO_3^-, and magnesium ions, Mg^{2+}.

(b) Describe the importance of the ions given in part (a).   **(3 marks)**

Nitrate ions are needed for protein synthesis, and magnesium ions are needed to make chlorophyll. Plants grow poorly if they cannot absorb enough of these ions.

You can revise the absorption of mineral ions by root hair cells on pages 5 and 15.

Now try this

There are laboratories that carry out disease identification for farmers. They require a sample of plant material and a sample of the soil in which the plant was growing. Describe **three** reasons why the laboratories need these samples.   **(3 marks)**

Remember that there are chemical tests for certain ions.

- Mg for chlorophyll
- Nitrogen for amino acid

Plant defences

Physical defences

Plants have physical defence responses to resist invasion by **microorganisms**.

1 **Cellulose cell walls**
The tough **cell walls** make it difficult for microorganisms to pass through. If a part of a plant is broken or **aphids** penetrate the **phloem** tissue, microorganisms can more easily infect the plant.

2 **Waxy leaf cuticle**
The upper and lower leaf epidermises are covered by a tough waxy **cuticle**. This layer reduces water loss from the leaf but also protects against microorganisms.

3 **Layers of dead cells**
Tree **bark** and layers of dead cells around stems form barriers against microorganisms. When bark falls off, the microorganisms fall off with it.

Chemical defences

Plants have chemical defence responses, including:
- antibacterial substances
- toxins.

The table gives some examples of **antibacterial** substances made by plants.

Plant	Substance
chilli pepper	capsaicin
garlic, onion	allicin
thyme	thymol

Do not try to learn these examples, but you will see that that they include smelly or irritating substances.

Plants make **toxins** (poisons) to deter **herbivores** from eating them. For example, foxgloves produce a toxin that affects the heart rate of herbivores that try to eat them. This toxin, digitalis, is used as a treatment for irregular heartbeat in humans (see page 41).

Worked example

Some cacti have hairs and thorns on them.

Give a reason why they have these features. **(1 mark)**

Hairs and thorns make it painful or unpleasant for herbivores that might try to eat the plant.

These are mechanical adaptations. Some plants have leaves that droop or curl when touched. This may scare away some animals, or cause insect pests to fall off.

Mimicry

Mimicry is a situation where one species is adapted to resemble another species in some way. Some plants use mimicry to trick animals. For example, the leaves of some species of passion flowers develop parts that look like butterfly eggs:
- Butterflies tend to avoid laying eggs where there are eggs already.
- Plants using mimicry reduce the chance of being eaten by caterpillars, because there are fewer eggs on them to hatch.

Mimicry like this is a **mechanical adaptation** to reduce the risk of damage by herbivores.

Now try this

1 Give **one** mechanical defence adaptation of plants. **(1 mark)**

2 Describe **three** ways in which physical barriers can protect a plant from pests and pathogens. **(3 marks)**

3 Explain the advantage to plants of producing substances that deter pests or kill pathogens. **(2 marks)**

Extended response – Infection and response

There will be at least one 6-mark question on your exam paper. For these questions, you will need to think scientifically and structure your answer logically, showing how the points you make are related to each other.

For the questions on this page, you can revise new medicines on pages 39 and 41.

Worked example

Schistosomiasis is a tropical disease caused by a parasitic worm. It leads to long-term disability, liver damage and even death. Over 60 million people each year are treated for the disease.

A single dose of a drug called praziquantel results in the death of the worms in the body. It is cheap and its side effects are mainly due to dying worms. However, praziquantel-resistant strains of the worm are appearing, so scientists are researching new drugs. Explain how a new drug to treat the disease must be tested before it is licensed for use. **(6 marks)**

The drug must first be tested on cells, tissues and live laboratory animals. This is to work out how toxic it might be, whether it works and what a suitable dose might be.

If the drug passes the first stages, it is then tested at very low doses using healthy volunteers and patients. This is to check whether it is safe to use in people, and to find out if it has any side effects.

A drug that passes these stages is then further tested in patients to find its optimum dose. Double blind trials are used to ensure valid results and to avoid false claims of success. Patients are randomly put into one of two groups. One group receives the drug. The other group receives a placebo which does not contain the drug. The groups are only revealed at the end of the trial so the results can be analysed.

These parasitic worms are not viruses, bacteria, protists or fungi. However, you should be able to apply your knowledge and understanding to different situations using given information.

Praziquantel has desirable features, so the emergence of resistant strains will cause concern amongst health professionals.

Command word: Explain

When you are asked to **explain** something, you must include reasoning or justification of the points you make.

This part of the answer describes preclinical testing and gives reasons for doing it.

This part of the answer describes clinical trials and gives reasons for doing them.

Double blind trials are explained here. Note that, for ethical reasons, people in the placebo group would be given praziquantel afterwards to treat their infection.

Now try this

The MMR vaccine protects against measles, mumps and rubella. The graph shows the percentage of young children given the MMR vaccine in England. The World Health Organization recommends that at least 95% should be immunised. Explain the importance of this target. **(6 marks)**

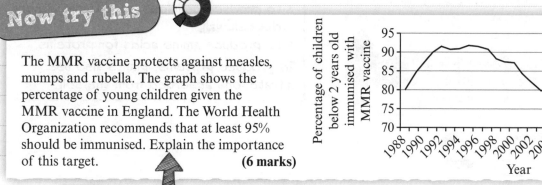

You can revise vaccination on page 39.

Photosynthesis

Plants make their own food by **photosynthesis**.

level 9 → (Not ~~made~~)

Reactants and products

Photosynthesis is an **endothermic** reaction – energy is transferred to the chloroplasts.

light

| carbon dioxide | + | water | → | glucose | + | oxygen |

$$6CO_2 \quad + \quad 6H_2O \quad \xrightarrow{Chlorophyll} \quad C_6H_{12}O_6 \quad + \quad 6O_2$$

diffuses into   absorbed by   used   diffuses from cells
cells from air   root hair cells, by the   into the air
                   transported in xylem plant

Rate of photosynthesis

The rate of photosynthesis depends upon **four** factors:
- temperature
- carbon dioxide concentration
- light intensity
- amount of chlorophyll.

The graph shows how the rate of photosynthesis varies with **temperature**.

If the temperature is too high, enzymes start to denature and the rate of photosynthesis slows down.

The graph shows how the rate of photosynthesis varies with **carbon dioxide concentration**. The shape of this graph:
- is different from the temperature graph
- is the same for **light intensity** and **amount of chlorophyll**.

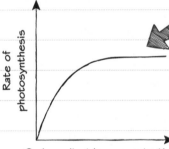

The rate of photosynthesis does not increase further when the carbon dioxide concentration becomes very high.

Worked example

Describe how a leaf may be tested for starch. **(4 marks)**

Heat the leaf for a minute in boiling water. Transfer the leaf to a test tube of ethanol, and heat in a hot water bath for a few minutes. Rinse the leaf with cold water and spread it on a white tile. Add a few drops of iodine solution. The leaf turns blue-black where starch is present.

Boiling water kills the leaf cells and softens their cell walls so iodine solution can enter. Ethanol removes the chlorophyll.

Uses of glucose

Glucose may be converted into insoluble starch for storage. Glucose may be used:
- for **respiration**
- to produce fat or oil for storage
- to produce cellulose, which strengthens the cell wall
- to produce amino acids for proteins.

To produce proteins, plants also use nitrate ions absorbed from the soil.

Now try this

Variegated leaves have white patches or stripes inbetween green areas. A variegated leaf is starch tested as described in the Worked example. Explain why the leaf only turns blue-black where the leaf was green.

(4 marks)

Limiting factors

Limits to rate

Factors affecting the rate of photosynthesis include:
- temperature
- carbon dioxide concentration
- light intensity
- amount of chlorophyll.

Any one of these may be a **limiting factor** – one that stops photosynthesis going faster.

Commercial growers need an understanding of limiting factors. They want to:
- enhance growing conditions to achieve the maximum photosynthesis rate
- still make a profit.

It costs money to provide heating, extra carbon dioxide and artificial lighting.

Determining rate

Pondweed provides an easy way to measure the rate of photosynthesis. The greater the rate of bubbling, the greater the rate of photosynthesis.

Temperature

Once the graph is horizontal, increasing the temperature no longer increases the photosynthesis rate. Remember that enzymes are **denatured** at high temperatures.

> Increasing the temperature here has a smaller effect than increasing the carbon dioxide concentration.

Three factors

The graph shows what happens when the light intensity is increased at different temperatures and carbon dioxide levels.

Comparing lines A and B:
- carbon dioxide is the limiting factor.

Comparing lines B and C:
- temperature is the limiting factor.

Worked example

The graph shows the rate of photosynthesis of tomato plants in a greenhouse under different conditions. Explain which factor shown is the limiting factor at point X. **(2 marks)**

Light intensity is the limiting factor here. Increasing the carbon dioxide concentration does not increase the rate, but increasing the light intensity does.

Now try this

Explain which factor is the limiting factor at point Y in the Worked example. **(2 marks)**

Investigating photosynthesis

 Practical skills You can investigate the effect of **light intensity** on the **rate of photosynthesis** in an aquatic plant.

Core practical

Investigating the effect of light intensity

Aim

to investigate the effect of light intensity on the rate of photosynthesis in pondweed

Apparatus

- boiling tube and test-tube rack
- bright lamp, ruler, stop watch
- sodium hydrogen carbonate solution
- freshly cut piece of pondweed

Method

1 Put the boiling tube in the test-tube rack.
2 Add sodium hydrogen carbonate solution.
3 Lower the piece of pondweed into the boiling tube.
4 Position the boiling tube 10 cm away from the lamp, and leave it for 5 minutes.
5 Count the number of bubbles produced in 1 minute. Repeat this twice more.
6 Repeat steps 4 and 5 at different distances

Results

Record your results in a suitable table. This one shows some sample mean results.

Distance (cm)	Bubbles per minute			
	Run 1	Run 2	Run 3	Mean
10				50
20				35
30				30
40				27

Analysis

1 For each distance, calculate the mean rate of bubbling, and the light intensity.
2 Plot a graph with:
  - mean rate of bubbling on the vertical axis
  - light intensity on the horizontal axis
3 Describe how the rate of bubbling changes as the light intensity increases.

> Pondweed such as *Elodea* is suitable to use. You can also use algae.

> Sodium hydrogen carbonate solution, $NaHCO_3$, is a source of carbon dioxide. It is included so that the carbon dioxide concentration is not a limiting factor.

🔢 Maths skills **Inverse square law**

The light intensity is inversely proportional to the square of the distance:

$$\text{light intensity} \propto \frac{1}{d^2}$$

where d = distance from lamp to pondweed.

You can convert all the distances in metres into equivalent light intensities. For example:

$d = 50\,cm$

$d = 50/100 = 0.50\,m$

$$\text{light intensity} = \frac{1}{0.50^2} = \frac{1}{0.25} = 4$$

You don't need any units if you do the same calculation for all the distances.

> Repeating the count at each distance allows you to check whether the results are reliable. You can do more repeats if, for example, you obtain very different results at a particular distance. The most reliable results are identical or very similar to each other.

> The Maths skills box describes how to do this.

> Choose sensible scales that allow the plotted points to occupy at least 50% of the area of the graph. Draw a curve of best fit.

Now try this

1 Express all the distances in the table in metres, m. Use these values to calculate the light intensity at each distance. Give your final answers to 2 significant figures. **(3 marks)**
2 Draw a suitable graph to show the results. Describe what the results show. **(5 marks)**

Respiration

Respiration in cells can happen **aerobically** (using oxygen) or **anaerobically** (without oxygen).

Respiration compared ⌐ Exothermic ⌐

Aerobic respiration	Anaerobic respiration
glucose + oxygen → carbon dioxide + water $C_6H_{12}O_6 + 6O_2 \rightarrow 6CO_2 + 6H_2O$	in muscles: glucose → lactic acid $C_6H_{12}O_6 \rightarrow 2C_3H_6O_3$
uses oxygen	does not use oxygen
complete oxidation of glucose	incomplete oxidation of glucose
exothermic reaction	exothermic reaction

Aerobic respiration transfers much more energy than anaerobic respiration.

Uses of energy

Cellular respiration is an exothermic reaction which happens continuously in living cells. It supplies all the energy needed for the processes that happen in living organisms.
These include:
- chemical reactions to build larger molecules
- movement
- active transport
- keeping warm.

Fermentation

Anaerobic respiration can also happen in plant cells and yeast cells. The overall reaction is different from the reaction in muscles:

glucose → ethanol + carbon dioxide
$C_6H_{12}O_6 \rightarrow 2C_2H_5OH + 2CO_2$

Anaerobic respiration in yeast cells is called **fermentation**. This reaction is useful in:
- making bread – the gas makes dough rise
- making alcoholic drinks such as wine and beer.

Respiration vs photosynthesis

The overall reaction for aerobic respiration is the reverse of the overall reaction for photosynthesis. The table shows some other differences between the two processes.

Aerobic respiration	Photosynthesis
exothermic	endothermic
gives out energy	takes in light energy
uses oxygen	produces oxygen
produces carbon dioxide	uses carbon dioxide
happens in mitochondria	happens in chloroplasts
happens in animal cells and plant cells	happens in plant cells

Worked example

Compare aerobic respiration with anaerobic respiration in muscles. **(2 marks)**

Both processes use glucose. Aerobic respiration needs oxygen but anaerobic respiration does not.

Glucose is not completely oxidised in anaerobic respiration, so it releases much less energy than aerobic respiration does.

Aerobic respiration produces two products, carbon dioxide and water. Anaerobic respiration produces just one product, lactic acid.

Remember that anaerobic respiration in muscles differs from that in plants and yeast.

Now try this

Use the information on this page to give **one** advantage and **one** disadvantage of:
(a) aerobic respiration; (b) anaerobic respiration.

(4 marks)

Responding to exercise

The human body reacts to the increased demand for energy during exercise.

Effects of exercise

When you exercise, your muscle cells respire faster. They need more oxygen and glucose, and release more carbon dioxide. So, during exercise, the following increase:
• heart rate
• breathing rate
• breath volume.

A greater heart rate means that blood is pumped faster around the body. It can transport oxygen and glucose to cells faster, and remove carbon dioxide faster.

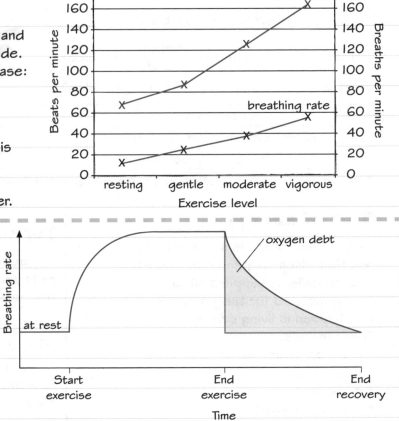

Oxygen debt

During exercise, insufficient oxygen may be supplied to the respiring muscle cells for aerobic respiration. When this happens, anaerobic respiration occurs in muscles:
• Glucose undergoes **incomplete oxidation**.
• **Lactic acid** builds up in muscles.
• The muscles become fatigued and stop contracting efficiently.

Blood transports lactic acid from the muscles to the **liver**, where it is converted back into glucose.

Oxygen debt is the amount of extra oxygen the body needs after exercise to react with the accumulated lactic acid and remove it from the cells. This is why, for a while after exercise, you still breathe heavily and your heart continues to beat faster than normal.

Worked example

Explain why the breathing rate and breath volume increase during exercise. **(3 marks)**

A greater breathing rate means that air is inhaled and exhaled more frequently, and a greater breath volume means that more air is moved in and out each time. The rate of respiration increases during exercise, and these changes allow oxygen to be taken into the body faster and waste carbon dioxide to be released faster.

The resting breathing rate is often counted over periods of one minute. During exercise, and after it, shorter periods of 15s or 30s may be better.

Breath volume is measured using a device called a spirometer.

Heart rate is measured by counting the **pulse rate** in the wrist. It is usually counted during periods of 15s, then multiplied by 4 to get the beats per minute.

Now try this

1 Explain why the muscles may become fatigued during exercise. **(2 marks)**

2 Explain why an oxygen debt develops during exercise. **(3 marks)**

Metabolism

Metabolism is the sum of all the reactions that happen in a cell or in the body.

Carbohydrates

Glucose is a simple sugar. Sugar molecules can react together in different ways to form:

- **glycogen**, for storage in the muscles and liver
- **starch**, for storage in plant cells
- **cellulose**, to strengthen plant cell walls.

Most animals cannot digest cellulose, but they can break down glycogen and starch to form glucose and other sugars.

digestion
amylase

carbohydrates, e.g. starch for energy storage in plants → sugars, e.g. glucose for respiration

synthesis

Lipids

Lipids are fats and oils. At room temperature:

- fats are solids
- oils are liquids.

A lipid molecule forms when a **glycerol** molecule reacts with three **fatty acid** molecules.

Lipids are used for:

- storage in plants and animals
- **insulation** against heat loss in animals.

digestion
lipase

lipids, e.g. for energy storage → fatty acids and glycerol, e.g. for respiration

synthesis

Proteins

Proteins include enzymes, and structural substances such as collagen and keratin.

Different **amino acids** join together in different ways to form proteins. You can revise this process on page 79.

digestion
protease

proteins, e.g. for muscle cells → amino acids, e.g. to make enzymes

synthesis

The liver

The **liver** has many functions. These include:

1 converting lactic acid, produced during anaerobic respiration, into glucose
You can revise this on page 50.

2 converting excess glucose into glycogen for storage
You can revise this on page 62.

3 converting excess amino acids to form ammonia (**deamination**), and converting this ammonia into urea for excretion
You can revise this on page 65.

Worked example

Describe how plants form proteins, starting with glucose produced by photosynthesis. **(3 marks)**

Plants absorb nitrate ions through their roots. Glucose and nitrate ions combine to form amino acids. These join end to end to form proteins.

Remember that energy transferred by respiration is used by organisms for continual processes, controlled by enzymes, that synthesise new substances.

The details of how amino acids are formed are complicated and involve many reactions.

Now try this

1 Give the meaning of 'metabolism'. **(1 mark)**

2 Name **three** types of digestive enzymes involved in metabolism. **(3 marks)**

Extended response – Bioenergetics

There will be one at least one 6-mark question on your exam paper. For these questions, you will need to think scientifically and structure your answer logically, showing how the points you make are related to each other.

For the questions on this page, you can revise responding to exercise on page 50.

Worked example

Explain the effects of temperature, light intensity, carbon dioxide concentration and the amount of chlorophyll on the rate of photosynthesis. **(6 marks)**

Photosynthesis involves the reaction between carbon dioxide and water:

$$\text{carbon dioxide} + \text{water} \xrightarrow{\text{light}} \text{glucose} + \text{oxygen}$$

As carbon dioxide is one of the reactants, its concentration in the air affects the rate of photosynthesis.
As the concentration increases, the rate of photosynthesis increases until it reaches a maximum rate determined by another factor.

Rate of photosynthesis vs Carbon dioxide concentration

Photosynthesis is an endothermic reaction in which energy is transferred to the chloroplasts by light. So the greater the light intensity, the greater the rate of photosynthesis. Chlorophyll in chloroplasts absorbs light, so the more chlorophyll there is, the greater the rate of photosynthesis. Graphs of rate against light intensity or amount of chlorophyll are similar in shape to the one sketched above.

The rate of chemical reactions increases as the temperature increases. So the rate of photosynthesis increases as the temperature increases. However, enzymes are involved in photosynthesis. They become denatured at high temperatures, so the rate of photosynthesis decreases above a certain temperature.

Command word: Explain

When you are asked to **explain** something, you must include reasoning or justification of the points you make.

Showing the equation helps to explain why carbon dioxide and light are important in photosynthesis.

If you sketch a graph, it may help save some writing and make your answer clearer.

This part of the answer links the absorption of light to the rate of photosynthesis. There is no need to sketch a graph since one is already given, and its relevance is explained here.

The answer could include that an enzyme becomes denatured when the shape of its active site changes irreversibly.

Now try this

Explain why breathing rate and heart rate increase during exercise. **(6 marks)**

Include aerobic respiration and anaerobic respiration in your answer (you can revise these on pages 49 and 50).

Homeostasis

Homeostasis involves the regulation of the internal conditions of a cell or organism.

Optimal conditions

Enzymes become **denatured** and no longer work if the temperature becomes too high, or if the pH becomes too high or low. The **optimum temperature** and **optimum pH** are the conditions in which an enzyme works best.

Homeostasis:

- maintains optimal conditions for enzyme action and all cell functions
- involves responses to changes in the internal or external conditions. (stimulus)

Control in the body

Many conditions are controlled in the human body. They include: ④ Salts (minerals)

1 body temperature
You can revise thermoregulation on page 60.

2 blood glucose concentration
You can revise this on page 62.

3 water levels
You can revise osmoregulation on page 64.

Control systems

Homeostasis uses automatic control systems. These may include:

- nervous responses (involving the <u>nervous system</u>)
- chemical responses (involving <u>hormones</u>).

All control systems have three main parts.

The diagram below outlines how these work to control the amount of water in the body.

1 **Receptors**. These are cells that **detect stimuli** (changes in the environment).

2 **Coordination centres,** such as the brain, spinal cord and pancreas. These receive information from receptors and process it.

3 **Effectors**, such as muscles or glands. These bring about the responses needed.

Worked example

Describe why the control mechanisms in homeostasis may be described as 'negative feedback mechanisms'. **(2 marks)**

When a change happens, the control mechanisms act to reverse the change. For example, if the water content increases, the control mechanism decreases it again.

If the water content decreases, the control mechanism increases it again.

Homeostasis also works to bring body temperature and blood glucose levels up or down to their optimal values.

Now try this

Explain what is meant by 'homeostasis'.

(3 marks)

Neurones

The nervous system allows humans to react to their surroundings and to coordinate their behaviour. **Neurones** are cells in the nervous system.

Types of neurones

There are three main types of neurones:

- **Sensory neurones** carry electrical impulses from **receptors** to the **central nervous system** (CNS).
- **Relay neurones**, found in the **CNS**, carry impulses from sensory neurones to motor neurones.
- **Motor neurones** carry impulses from the CNS to **effectors**, such as muscles and glands.

CNS

Sensory neurone

The electrical impulse jumps from one gap in the myelin sheath to the next, speeding up the rate of transmission.

skin receptor cells   direction of impulse   cell body   Axon terminals pass impulses to other neurones.

dendron   **axon**   fatty **myelin sheath**

Dendrites collect impulses from receptor cells.

carry impulses long distances

This insulates the neurone. The electrical impulse cannot cross the fatty myelin sheath.

Relay neurone

Dendrites collect impulses from sensory neurones.

cell body

axon

Axon terminals pass impulses to other neurones.

Worked example

myelin sheath   axon   cell body   axon terminals   dendrite

The diagram shows the structure of a motor neurone. Explain how this cell is adapted to its function in the nervous system.   **(4 marks)**

The axon is long so it can transmit impulses over long distances. The dendrites can receive impulses from other neurones. The axon terminals can connect with effectors such as muscles and glands.

The fatty myelin sheath insulates the axon from other neurones, and the electrical impulses can jump quickly from one gap to another.

Nerve impulses travel from **dendrites** towards the **cell body**, and from the cell body towards the **axon terminals**.

All types of neurone have an **axon**, but:
- motor neurones and relay neurones do not have a **dendron**.

The insulation provided by the myelin sheath stops impulses travelling randomly into neighbouring neurones.

Now try this

1 Compare the roles of sensory, relay and motor neurones in the nervous system.   **(3 marks)**

2 Explain how the structure of a sensory neurone is related to its function.   **(4 marks)**

Reflex actions

The **central nervous system**, CNS, is the brain and spinal cord.

Coordinating responses

The CNS coordinates the response of effectors:

organ

change in environment → stimulus → receptor → (sensory neurone) → CNS / coordinator → (motor neurone) → muscle or gland / effector → contraction or hormone / response

ถ้า

Reflex actions do not involve the conscious part of the brain. This means that they are:
• **automatic** (you do not have to think for them to happen)
• **rapid**.

A **reflex arc** works like this:

CNS

stimulus → receptor → sensory neurone → relay neurone → motor neurone → effector → response

Reflex actions are **innate** – they are not learned responses.

They are important because they:
• control basic functions such as breathing
• help to avoid danger and harm.

Synapses

The point where two neurones meet is a **synapse**. Electrical nerve impulses cannot cross the gap in a synapse, so the signal is transmitted by **neurotransmitters**. This process:
👍 works in one direction only
👍 allows one nerve impulse to produce impulses in several other neurones ✳
👎 is slower than nerve impulses.

(1) Electrical nerve impulse reaches end of axon.

synapse

(2) Electrical impulse causes chemical **neurotransmitter** to be released into gap between neurones.

(3) Neurotransmitter causes new electrical impulse in next neurone.

ถ้า

Worked example

The diagram shows structures involved in a reflex arc.
(a) Name the structures labelled A, B and C in the diagram.
A – sensory neurone   B – relay neurone   **(3 marks)**
C – motor neurone
(b) Explain how the reflex shown helps to increase the chance of survival. **(2 marks)**
A hot flame or surface would damage the skin. The reflex allows the hand to move away quickly enough, without conscious thought, to reduce the chance of being injured.

A —
spinal cord
C —
B
biceps muscle
heat receptor in the skin

Now try this

Relay neurones also send impulses to neurones that lead to the brain. Explain why you only become aware of a reflex action after it has happened.
(2 marks)

Investigating reaction times

Practical skills You can investigate the effect of a factor on human **reaction time**.

Core practical

Effect of practice on reaction times

Aim

to investigate the effect of practice on human reaction times

> You could also plan and carry out an investigation to see if exercise or having a cola drink containing caffeine has an effect.

Apparatus

- ruler, 1 m long, marked in cm • a partner to work with

> You may be given a ruler with a paper scale marked in reaction times, or a table of data so you can convert readings in cm into reaction times.

Method

1 Sit on a stool with the forearm of your weaker hand on the bench. Your hand should hang over the edge.

2 Open your thumb and forefinger. The ruler will fall between them and you will need to catch it with your thumb and forefinger.

> If you are right handed, your weaker hand is your left hand, and vice versa.

3 Your partner should then hold the ruler upright so the 0 cm mark is level with the top of your thumb.

> Sit up straight, and focus across the lab rather than on the ruler itself.

4 Without telling you, your partner should let go of the ruler. Catch the ruler as soon as you can, using your thumb and forefinger.

5 Record the reading on the ruler, level with the top of your thumb, to the nearest cm.

> The ruler should lightly touch your thumb so you can feel when it starts to drop.

6 Repeat steps 2–5 ten times.

7 Swap roles and repeat steps 1–6.

Results

Record your results in a suitable table.

Drop number	Reading (cm)	Reaction time (ms)
1		
2		

Analysis

1 For each reading, determine the equivalent reaction time using a data table or by calculation.

2 Plot a bar chart with:
   - reaction time on the vertical axis
   - drop number on the horizontal axis.

3 Describe what the results show.

Maths skills — Calculating reaction time

This ruler falls at a constant acceleration because of gravity. This means you can convert the reading in cm into time in milliseconds, ms:

$$\text{reaction time in ms} = \sqrt{2000 \times \text{cm}}$$

For example, if the ruler falls 75 cm:

$$\text{reaction time} = \sqrt{2000 \times 75}$$

$$= \sqrt{150\,000}$$

$$= 387\,\text{ms}$$

Most readings will be to 2 significant figures, so you only need to give the final answer to 2 significant figures. In this example, 390 ms.

Now try this

Explain why you should not do any practice before starting the investigation. **(2 marks)**

The brain

The **brain** is made of billions of interconnected neurones and controls complex behaviour.

Brain structure

The brain has different regions that carry out different functions.

The cerebral cortex controls **voluntary movement, interprets** sensory information and is responsible for **learning** and **memory**.

Contex

cerebral cortex

The cerebellum coordinates and controls precise and smooth **movement**.

The medulla regulates the **heart beat**, **breathing** and other unconscious processes.

medusa

medulla

cerebellum

→ *balance*

Investigating brain function

The brain is very delicate and complex. This makes it difficult to investigate the brain.

Scientists who study the structure and function of the brain have mapped different regions to particular functions by:

1 Studying people with brain damage. Damage to different regions of the brain causes different changes in body function.

2 Using electrical stimulation. Electrodes are used to pass electrical impulses to different regions of the brain, and the effects of this are studied.

3 Using MRI scanning. MRI scanners can build up detailed images of the brain and other body tissues.

HT

MRI scanning

Magnetic resonance imaging, **MRI**, uses radio waves and very strong magnetic fields. It does not have harmful side effects, but patients must remove all magnetic materials before a scan.

damaged area of the brain

Worked example

Describe two problems in treating brain damage or disease. **(2 marks)**

The brain is very delicate, so a lot of care is needed during surgery to avoid further damage. The brain is also very complex, so it may be difficult to identify the exact area to treat.

The brain itself does not have pain receptors, so patients can be conscious during surgery.

The **blood–brain barrier** only allows certain substances to reach the brain, so medical drugs do not always reach it.

Now try this

Name the part of the brain responsible for each of the following activities:
(a) increasing heart rate during exercise **(1 mark)**
(b) recognising a person from a photograph. **(1 mark)**

The eye

The eye is a sense organ that contains receptors sensitive to light intensity and colour.

Structure of the eye

suspensory ligaments
iris
lens
cornea
ciliary muscles

retina
sclera
optic nerve

Light travels into the eye through:
- the tough, transparent **cornea**
- the pupil (the hole made by the **iris**)
- the transparent **lens**.

Light reaches the receptor cells in the **retina**:
1. **Rods** are sensitive to low light levels and respond to how light or dark an object is.
2. **Cones** are sensitive to bright light and respond to different colours.

Structure	Features	Function
sclera	tough, white outer layer	• protects against damage • provides attachment for eye muscles
retina	contains receptor cells (rods and cones)	• detects light • sends impulses to the optic nerve
optic nerve	contains sensory neurones	• sends impulses to the optical centre in the brain

Focusing light

Accommodation is the process of changing the shape of the lens to focus on objects.

	Focusing on a near object	Focusing on a distant object
Light rays	rays of light from a near object eye lens ใกล้	rays of light from a distant object eye lens ไกล
Ciliary muscles	contract	relax
Suspensory ligaments	loosen	pulled tight
Lens	thicker refracts light strongly	pulled thin refracts light slightly

Worked example

The diagrams show how the size of the pupil changes when the light becomes dimmer. Describe how this change occurs. **(3 marks)**

The iris contains muscles. In dim light, its circular muscles relax and its radial muscles contract. The pupil dilates (becomes larger). This lets more light into the eye.

bright light   dim light

This is an example of a reflex action. You can revise these on page 55.

Now try this

Describe the change to the pupil when we go into bright light, and how it occurs. **(3 marks)**

Eye defects

Myopia (short sightedness) and **hyperopia** (long sightedness) are common eye defects.

Myopia → มไม-โอ (สายตา สั้น)

A person with **short sightedness**:
- can see near objects clearly
- but distant objects appear blurred.

Light focuses in front of the retina because:
- the eyeball is too long, or *too thick*
- the cornea is too curved.

Hyperopia → Hype (คนเป็นสายตายาวเยอะ)

A person with **long sightedness**:
- can see distant objects clearly
- but near objects appear blurred.

Light focuses behind the retina because:
- the eyeball is too short, or *to thin*
- the cornea is too flat.

short sight

long sight

This defect is usually treated with spectacles that have **concave lenses**. These **refract** light so it **diverges** before reaching the eye. This means that light focuses onto the retina.

concave lens (diverging)

This defect is usually treated with spectacles that have **convex lenses**. These refract light so it **converges** before reaching the eye. This means that light focuses onto the retina.

convex lens (converging)

Other treatments

Other treatments for myopia or hyperopia include:
- hard or soft **contact lenses**
- **laser eye surgery**, in which a laser is used to change the cornea's shape so it refracts light differently before reaching the lens
- surgically implanting an **artificial lens**, either as a replacement for the natural lens or to work with it.

Cataracts

A **cataract** occurs when a part of the lens becomes less transparent. This stops some of the light reaching the retina, so vision becomes cloudy or blurred. Cataracts are treated by surgery:
- The natural lens is removed.
- A polymer lens is inserted.

The artificial lens has a fixed shape, so spectacles may still be needed afterwards.

Worked example

Explain why the incidence of hyperopia tends to increase with age. **(3 marks)**

The lens becomes harder as people age. This makes it increasingly difficult for the ciliary muscles to change the lens shape sufficiently. Eventually the muscle cannot make the lens thick enough to focus on near objects.

Now try this

1 Compare the difference in vision between a person with myopia and a person with hyperopia. **(2 marks)**

2 Explain how spectacles are used to treat hyperopia. **(5 marks)**

Thermoregulation

Thermoregulation involves several processes that keep body temperature close to 37°C.

The skin

The **skin** is an organ that plays a major role in thermoregulation.

releases sweat when warm to lose heat by evaporation

Vasodilation/vasoconstriction changes blood flow through surface capillaries depending on temperature.

hair

hair muscle

sweat gland

epidermis

blood capillary

blood vessel

dermis

Cooling down

If the body temperature is too high:

1. **Vasodilation** occurs.
   Blood vessels supplying the skin become wider (they **dilate**). More blood flows through skin capillaries, so more energy is transferred to the environment by **heating**.

2. The **sweat glands** release sweat. Energy must be transferred to allow the continued evaporation of water in **sweat**. This causes a transfer of energy from the skin, cooling the body.

Staying warm

If the body temperature is too low:

1. **Vasoconstriction** occurs.
   Blood vessels supplying the skin become narrower (they **constrict**). Less blood flows through skin capillaries, so less energy is transferred to the environment by heating.

2. The sweat glands stop releasing sweat.

3. **Skeletal muscles** contract and relax repeatedly. This **shivering** releases energy by heating because of increased respiration in the muscle cells.

Monitoring and control

The **thermoregulatory centre** in the brain monitors and controls body temperature.

Thermoregulatory centre contains receptors sensitive to blood temperature.

nervous impulses sent to brain

temperature receptors in skin

Worked example

Explain how the hairs on the skin play a role in regulating body temperature.   **(3 marks)**

If the body temperature becomes too low, the hair muscles contract, pulling the skin hairs upright. This traps a thicker layer of insulating air close to the skin, which helps to reduce heat losses from the skin.

This is why you get 'goosebumps' in the cold. The skin hairs lie back down again in warm conditions.

Now try this

Suggest why we need to drink more on a hot, sunny day.   **(2 marks)**

Hormones

The **endocrine system** comprises glands that secrete hormones directly into the bloodstream.

Secreting hormones

A **hormone** is a substance that:
- is secreted by a **gland**
- is transported in the bloodstream
- has an effect on a **target organ**.

The **pituitary gland** is a 'master gland'. It secretes several hormones in response to body conditions. These hormones then act on other glands, stimulating the release of other hormones.

Hormones and nerves

The endocrine system and the nervous system are both involved in responses to changes in the environment. However, there are differences in their effects and how quickly they work.

	Hormones	Nerve impulses
Speed	slow	fast
Duration	long	short

Glands and hormones

Different endocrine glands produce different hormones.

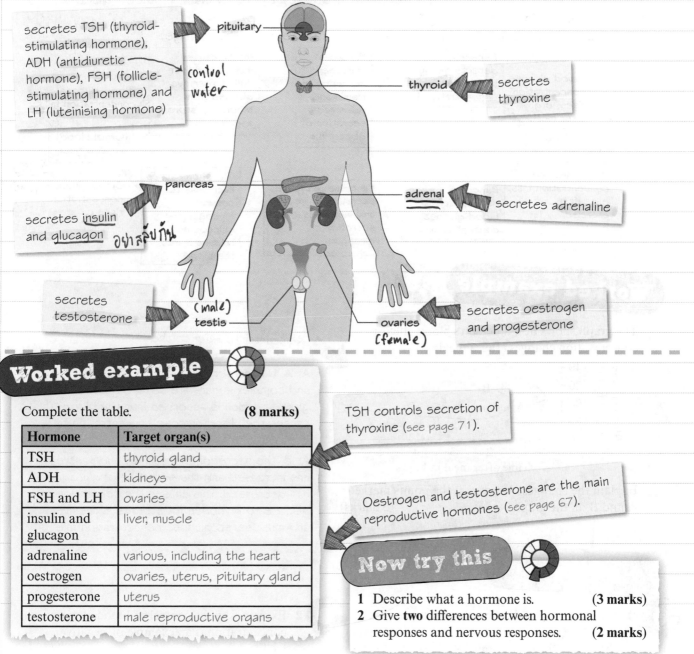

secretes TSH (thyroid-stimulating hormone), ADH (antidiuretic hormone), FSH (follicle-stimulating hormone) and LH (luteinising hormone)

control water

pituitary

thyroid ← secretes thyroxine

pancreas

secretes insulin and glucagon อย่าสลับกัน

adrenal ← secretes adrenaline

secretes testosterone → (male) testis

ovaries (female) ← secretes oestrogen and progesterone

Worked example

Complete the table.   **(8 marks)**

TSH controls secretion of thyroxine (see page 71).

Hormone	Target organ(s)
TSH	thyroid gland
ADH	kidneys
FSH and LH	ovaries
insulin and glucagon	liver, muscle
adrenaline	various, including the heart
oestrogen	ovaries, uterus, pituitary gland
progesterone	uterus
testosterone	male reproductive organs

Oestrogen and testosterone are the main reproductive hormones (see page 67).

Now try this

1 Describe what a hormone is.   **(3 marks)**
2 Give **two** differences between hormonal responses and nervous responses.   **(2 marks)**

Blood glucose regulation

The pancreas monitors and controls blood glucose concentration.

Insulin and glucagon

Two hormones, **insulin** and **glucagon**, control blood glucose levels in the body. They have opposite effects on the body.

Sugar In Insulin	Glucagon
causes glucose → glycogen	causes glycogen → glucose
decreases blood glucose levels	increases blood glucose levels

new

Glycogen and starch

Glycogen and starch are complex molecules produced when many **glucose** molecules join together in different ways:
- Plants store starch in their cells.
- Animals store **glycogen** in their liver cells and muscle cells.

Don't confuse 'glycogen' with 'glucagon'.

Negative feedback cycle

Glucagon interacts with insulin in a **negative feedback cycle** to maintain blood glucose concentration within a small range:
- A change in concentration causes responses that reverse the change.

remember use "cause"

Blood glucose concentration rises, e.g. after meal.

normal blood glucose range

Blood glucose concentration falls, e.g. after exercise.

Pancreas detects rise in blood glucose concentration.

Pancreas increases secretion of insulin and decreases secretion of glucagon.

Insulin causes muscle and liver cells to remove glucose from blood and store it as glycogen.

Blood glucose concentration falls.

normal blood glucose range

Pancreas detects fall in blood glucose concentration.

Pancreas decreases secretion of insulin and increases secretion of glucagon.

Glucagon causes liver cells to convert glycogen to glucose and release it into the blood.

Blood glucose concentration rises.

 # goal
negative feedback

Worked example

The graph shows how a person's blood glucose concentration changes after a meal.

Explain the changes observed at the points marked A and B on the graph. **(5 marks)**

Remember to refer to the graph.

At A, the person has just had their meal. Starch and complex sugars are being digested to form glucose, which is absorbed into the blood. This causes the blood glucose concentration to increase over the first hour after eating.

At B, the secretion of insulin by the pancreas has increased and the secretion of glucagon has decreased. This causes the conversion of glucose to glycogen for storage in the liver and muscles, so glucose concentration falls.

Now try this

1 (a) Name the endocrine gland that secretes glucagon. **(1 mark)**
   (b) Name the target organ of glucagon. **(1 mark)**

2 Explain why the control of blood glucose concentration is an example of negative feedback. **(2 marks)**

Diabetes

Diabetes is a condition in which the body cannot properly control blood glucose concentration.

Two types of diabetes

There are two types of diabetes, with different causes and treatments. Type 2 diabetes is associated with excess body mass, but type 1 diabetes is not.

	Type 1 diabetes	Type 2 diabetes (obesity)
Cause	• Pancreatic cells are destroyed by the body's immune system. • The pancreas does not produce sufficient insulin.	• Liver cells and muscle cells (no longer respond) to insulin. • The pancreas may not produce enough insulin.
Treatment	insulin injections	a carbohydrate-controlled diet and exercise
Diagnosis	often in childhood	usually when over 30 years old

Body mass index

Obesity is a risk factor for type 2 diabetes. **Body mass index**, BMI, is one indicator of obesity. The table shows categories of BMI that are often used.

BMI (kg/m²)	Category
<18.5	underweight
18.5–25	normal weight
25–30	overweight
>30	obese

🖩 Maths skills — Calculating BMI

You can calculate a person's BMI using:

$$BMI = \frac{\text{mass in kg}}{(\text{height in m})^2}$$

For example, the BMI of a 1.73 m tall person of mass 66 kg is:

$$\frac{66}{(1.73)^2} = \frac{66}{3.0} = 22 \text{ kg/m}^2$$

This person is in the normal weight category.

Worked example

The incidence of type 2 diabetes was studied in 51 529 men and 114 281 women with different BMIs.

Body mass index in kg/m²

The people should also be matched for factors such as age to obtain valid results.

(a) Describe what the results show. **(4 marks)**

The risk of developing type 2 diabetes increases as BMI increases. The risk increases slightly in the normal weight categories (up to 24.9) but increases greatly in the obese categories (over 30). Women are more likely to have the condition than men. For example, twice as many women with the highest BMIs are affected than men.

(b) Describe the significance of the sample sizes in this study. **(2 marks)**

These are very large sample sizes, so the results are likely to be valid.

Now try this

Explain how the dose of insulin needed by a person with type 1 diabetes will vary, depending on the food eaten and amount of exercise taken.

(3 marks)

Controlling water balance

The **urinary system** maintains water balance, and removes waste products.

Water loss

Body cells do not work efficiently if they lose or gain too much water by **osmosis**:

- Water leaves the body from the **lungs** when we exhale.
- Water, ions and urea are lost from the body in **sweat**.

The body cannot control the loss of water, ions or urea by the lungs or the skin.

Control is achieved by the **kidneys** in the urinary system. Kidneys produce urine by:

- filtration of the blood
- **selective reabsorption** of water and other useful substances.

Osmotic changes and cells

Cells are affected by changes in concentration of body fluids. The effects on red blood cells are easily observed with a light microscope.

placed in high concentration → loses water → becomes crinkled

placed in low concentration → gains water → eventually bursts

You can revise osmosis on page 13.   CUS no wall

Structure of the urinary system

Make sure you know the difference between urea and urine, and between ureter and urethra.

Urea is produced from the breakdown of excess amino acids in the **liver**. It is toxic in excess.

The **renal veins** carry cleaned blood back to the body.

The **ureters** carry **urine** from the kidneys to the bladder.

The **bladder** stores urine.

Urine flows through the **urethra** to the outside of the body.

The **renal arteries** carry blood from the body to the kidneys.

The **kidneys** remove substances including urea from the blood and make urine.

A muscle keeps the exit from the bladder closed until we decide to urinate.

Worked example

The table shows the typical concentrations of substances in blood plasma and urine.

Substance	Concentration (g/dm³)	
	Plasma	Urine
urea	0.3	18
salts	9	20
glucose	1	0

Describe the differences between the composition of plasma and urine. **(4 marks)**

The concentration of urea in the urine is sixty times greater than its concentration in plasma. The concentration of salts is just over twice as much in the urine. No glucose is found in the urine, even though it is found in the plasma.

The concentrations in urine vary, depending upon how much water is reabsorbed.

After filtering the plasma, the kidneys reabsorb all the glucose and some of the salts and urea.

Now try this

Think about the size protein molecules.

Suggest a reason why the kidneys cannot filter proteins from the plasma. **(2 marks)**

remember

ADH and the kidneys

ADH <u>increases the permeability of the kidney tubules</u>, causing more water to be reabsorbed.

Water loss

Each kidney has around a million **nephrons**. These are the functional units where:

- blood is filtered
- various substances are reabsorbed
- urine is produced.

The **tubule** in a nephron is the tube where reabsorption happens. This includes enough water to maintain the body's water level.

> The structure of the nephron with its tubule is shown only so you can see what happens there more easily.

In a nephron

blood from renal artery

blood to renal vein

filtration of small molecules from blood

selective reabsorption of glucose

reabsorption of water

urine containing urea and excess water to ureter

Control of water content

The control of water reabsorption is an example of a **negative feedback** cycle:

- a change in water level causes responses that reverse the change
- so that water level is maintained within a small range.

You can revise negative feedback cycles on pages 53 and 62.

The brain senses there is not enough water in the blood.

The pituitary gland secretes more ADH.

<u>More ADH</u> makes the tubules more <u>permeable</u>.

So more water is reabsorbed from the kidney tubule back into the blood.

A small volume of concentrated urine is produced.

pituitary gland

The brain senses there is too much water in the blood.

The pituitary gland secretes less ADH.

Less ADH makes the tubules less permeable.

So less water is reabsorbed from the tubule back into the blood.

A large volume of dilute urine is produced.

Worked example

Explain how urea is produced in the body. **(3 marks)**

Excess amino acids are produced when dietary proteins are digested. These must be excreted safely. The liver forms ammonia from these excess amino acids. As ammonia is toxic, it is immediately converted to urea.

Amino acids, ammonia and urea all contain nitrogen atoms:

- The process of forming ammonia from amino acids is called **deamination**.
- Urea is a nitrogenous waste product (it contains nitrogen atoms).

Now try this

Alcohol inhibits the production of ADH. Explain the effects of alcoholic drinks on urine production.

(4 marks)

Kidney treatments

People with **kidney failure** may be treated by organ transplant or by using kidney dialysis.

Organ transplant

A healthy organ may be provided by a donor. It is **transplanted** by surgery into a patient.

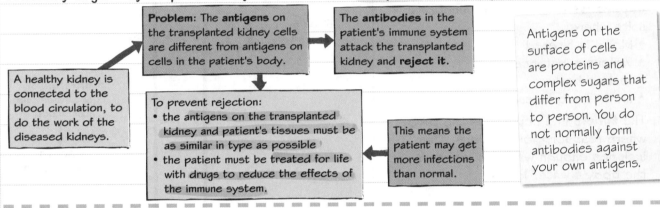

A healthy kidney is connected to the blood circulation, to do the work of the diseased kidneys.

Problem: The **antigens** on the transplanted kidney cells are different from antigens on cells in the patient's body.

The **antibodies** in the patient's immune system attack the transplanted kidney and **reject it**.

To prevent rejection:
- the antigens on the transplanted kidney and patient's tissues must be as similar in type as possible
- the patient must be treated for life with drugs to reduce the effects of the immune system.

This means the patient may get more infections than normal.

Antigens on the surface of cells are proteins and complex sugars that differ from person to person. You do not normally form antibodies against your own antigens.

Kidney dialysis

Kidney **dialysis** uses a machine to carry out the functions of the kidneys.

blood flow

dialysis fluid

Dialysis tubing is partially permeable.

Diffusion restores the normal concentrations of dissolved substances in the blood.

Urea diffuses out of the blood into the fluid.

Dialysis fluid contains the same concentration of useful substances as blood so glucose and useful mineral ions are not lost.

Feature	Reason
Dialysis tubing is partially permeable.	Blood cells, platelets and plasma proteins are not removed from the blood.
Dialysis fluid entering the machine has no urea.	This provides a large concentration difference so urea diffuses quickly from the blood.

Worked example

Compare the advantages and disadvantages of kidney dialysis and organ transplant as treatments for kidney failure. **(4 marks)**

A kidney transplant allows the patient to have a normal diet, free of visiting the hospital several times a week. However, donor kidneys are less readily available than dialysis machines, and carry a risk of organ rejection.

Dialysis patients must control the amount of protein in their diet to reduce the production of urea. A dialysis session takes a few hours.

Immunosuppressant drugs must be taken for life, to reduce the risk of the patient's immune system rejecting the transplanted kidney.

Now try this

Explain why, in terms of diffusion and the factors affecting it, fresh dialysis fluid is circulated through the dialysis machine. **(3 marks)**

Reproductive hormones

The menstrual cycle

Between puberty and about the age of 50, women have a **menstrual cycle** that repeats about every 28 days. During the cycle, changes take place in the ovaries and the uterus.

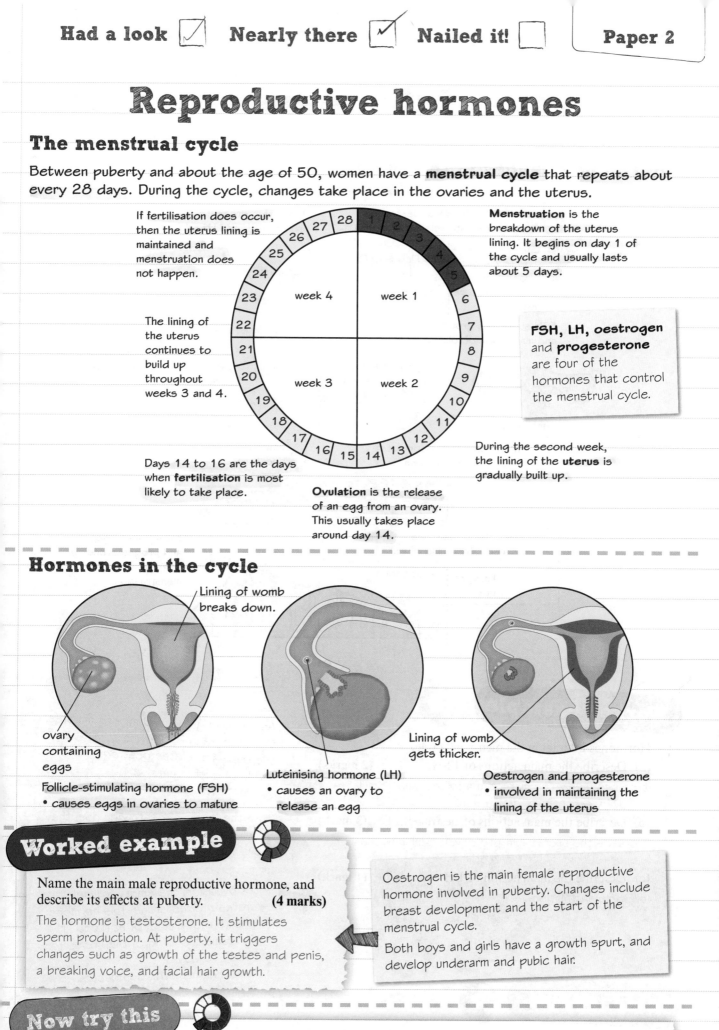

If fertilisation does occur, then the uterus lining is maintained and menstruation does not happen.

Menstruation is the breakdown of the uterus lining. It begins on day 1 of the cycle and usually lasts about 5 days.

The lining of the uterus continues to build up throughout weeks 3 and 4.

FSH, LH, oestrogen and **progesterone** are four of the hormones that control the menstrual cycle.

During the second week, the lining of the **uterus** is gradually built up.

Days 14 to 16 are the days when **fertilisation** is most likely to take place.

Ovulation is the release of an egg from an ovary. This usually takes place around day 14.

Hormones in the cycle

Lining of womb breaks down.

ovary containing eggs

Follicle-stimulating hormone (FSH)
• causes eggs in ovaries to mature

Luteinising hormone (LH)
• causes an ovary to release an egg

Lining of womb gets thicker.

Oestrogen and progesterone
• involved in maintaining the lining of the uterus

Worked example

Name the main male reproductive hormone, and describe its effects at puberty. **(4 marks)**

The hormone is testosterone. It stimulates sperm production. At puberty, it triggers changes such as growth of the testes and penis, a breaking voice, and facial hair growth.

Oestrogen is the main female reproductive hormone involved in puberty. Changes include breast development and the start of the menstrual cycle.

Both boys and girls have a growth spurt, and develop underarm and pubic hair.

Now try this

Explain why failure to menstruate may be the first sign of pregnancy. **(2 marks)**

Control of the menstrual cycle

FSH, LH, oestrogen and progesterone interact in their control of the menstrual cycle.

Hormone changes in the cycle

FSH and LH from the pituitary gland near the brain

High levels of oestrogen stimulate release of more LH.

Increasing progesterone inhibits FSH and LH release.

Low levels of progesterone allow FSH to be released.

blood levels of FSH

LH levels

FSH stimulates growth and maturation of follicles.

LH surge triggers ovulation.

ovary

growth of follicle

ovulation

yellow body

Yellow body forms then breaks down.

Maturing follicles stimulate oestrogen production.

Yellow body releases progesterone.

blood levels of oestrogen

progesterone levels

Increasing oestrogen causes thickening of wall.

Falling oestrogen and progesterone trigger menstruation.

lining of uterus

menstruation

14   28
Days

Worked example

Hormones interact in the menstrual cycle.

(a) Describe the main actions of FSH. **(2 marks)**

It causes an egg to mature in a follicle, and stimulates the ovary to produce oestrogen.

(b) Describe the main actions of oestrogen. **(3 mark)**

It causes the uterus lining to develop, inhibits the further release of FSH, and stimulates the release of LH.

(c) Describe the main action of LH. **(1 mark)**

It triggers ovulation.

As a follicle grows and an egg matures, the increase in FSH levels stimulates the ovary to release oestrogen, and the levels of oestrogen rise.

As the levels of oestrogen rise, they inhibit the further release of FSH. This ensures that only one egg matures.

After ovulation, the empty follicle secretes progesterone. This hormone:
• inhibits the release of FSH and LH
• maintains the lining of the uterus for about 10 days
• falls when menstruation occurs.

Now try this

Levels of progesterone remain high if an egg is fertilised. Explain the effects of this on the levels of FSH and LH, and on the lining of the uterus. **(4 marks)**

Contraception

Fertility can be controlled by hormonal and non-hormonal methods of contraception.

Oral contraceptives

Oral **contraceptives** may contain the hormones **oestrogen** and **progesterone**.

These hormones inhibit the release of <u>FSH</u>. This means that:

- eggs do not mature, so (~~as No~~ ~~Egg~~)
- no eggs are ready for release from an ovary.

Use of oral contraceptives carries a slight risk of side effects including:

- raised blood pressure
- blood clots (thrombosis) //
- breast cancer ✓

Worked example

Early oral contraceptives used much higher doses of oestrogen than modern ones. Some modern oral contraceptives contain only progesterone. Explain why the amounts and type of hormone were changed. **(2 marks)**

Large amounts of oestrogen caused side effects in women. The newer pills are still effective but cause fewer side effects.

Progesterone injections, implants or skin patches may allow slow release over months or years.

Surgical methods

Women are sterilised by blocking or sealing the oviducts that lead from the **ovaries** to the **uterus**. This stops eggs reaching sperm from a man and being fertilised. The **menstrual cycle** continues in women having this surgery, and they still have periods.

Men are sterilised by a **vasectomy** (see diagram). The sperm ducts are cut, stopping sperm getting into the semen.

Women and men must use other forms of contraception for a few months after surgery. The surgery is difficult to reverse.

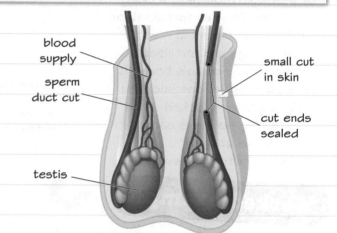

blood supply
sperm duct cut
small cut in skin
cut ends sealed
testis

Worked example

Describe two barrier methods of contraception.
(4 marks)

These rely on stopping sperm reaching an egg. A condom is a thin sheath made from latex. When placed over the penis, it collects semen. A diaphragm is made of thin latex. When it is placed over the woman's cervix, it stops sperm entering the uterus.

Both methods fail if they are used incorrectly or the latex is damaged. Their effectiveness increases when used with a **spermicide**, a substance that kills or disables sperm.

Other methods

An **intrauterine device**, IUD, is a small T-shaped plastic and copper object. It is inserted into the uterus by a doctor or nurse. An IUD:

- stops sperm and eggs from surviving in the uterus or oviducts
- may also prevent a fertilised egg from implanting in the uterus.

Normal fertility returns when a doctor or nurse removes it again.

Some couples practise **natural family planning**. They abstain from intercourse at the time in the menstrual cycle when an egg may be in the oviduct.

Now try this

Describe **two** ways in which a woman can temporarily control her fertility. **(4 marks)**

Treating infertility

Hormones are used as 'fertility drugs' to treat infertility, the inability to have a baby.

Fertility drugs

Doses of FSH and LH may help a woman with low FSH levels to become pregnant normally.

Artificial **FSH** is given to stimulate:
• an egg to mature into a follicle
• the production of oestrogen.

After giving FSH, artificial **LH** is given to:

• trigger ovulation. These two fertility drugs may then allow a woman to become pregnant naturally.

The doses are controlled to avoid the production of many mature eggs at once, which could lead to multiple births, such as triplets or quadruplets.

IVF treatment

An *in-vitro* process happens in a test tube, culture dish or elsewhere outside a living organism. *In-vitro* **fertilisation (IVF)** is a process in which mature eggs are fertilised outside the woman's body. It is used in situations such as:
• damaged or blocked oviducts, so eggs cannot reach the uterus
• few or malformed sperm.

1. FSH and LH are given to cause egg maturation and release.

ovary

2. Eggs are taken from the ovaries.

3. The eggs are mixed with sperm in a dish for fertilisation.

4. The fertilised eggs develop into embryos.

uterus (womb)

5. When the embryos are tiny balls of cells, one or two of them are placed in the woman's womb to develop.

Worked example

The table gives information about IVF treatments carried out in the UK in 2008.

Age of woman	Successful treatment (%)
<35	33.1
35–37	27.2
38–39	19.3
>39	10.7

Compare the benefits and problems that may arise from using IVF to treat infertility. **(3 marks)**

IVF helps some couples have a baby when they might not otherwise be able to. However, as the woman's age increases, the chance of having a baby decreases. Even the youngest women have only a 1 in 3 chance of becoming pregnant.

Drawbacks of fertility treatment

Fertility treatment gives a woman the chance to have a baby of her own. However, it:
• is emotionally stressful
• is physically stressful
• can lead to multiple births (a risk to the babies and to the mother)
• has low success rates.

The chances of success decrease very quickly with age over 35. The chance of successful treatment falls by 30% between ages 35–37 and ages 38–39. The chance of successful treatment falls by a further 55% after age 39.

Now try this

1 Describe **three** drawbacks of fertility treatment. **(3 marks)**

2 Explain why doses of hormones are carefully controlled in fertility treatment. **(3 marks)**

Adrenaline and thyroxine

Adrenaline and thyroxine are hormones that affect several target organs.

Adrenaline

Adrenaline is produced by the **adrenal glands** in times of fear or stress. It brings about changes that prepare the body for 'flight or fight'. The effects of adrenaline include:
- conversion of glycogen into glucose
- increases in heart rate and breathing rate.

These changes increase the delivery of oxygen and glucose to the brain and muscles.

Adrenal glands

These glands are at the top of the **kidneys**.

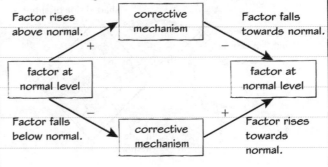

adrenal gland
kidney

Thyroxine

Thyroxine is produced by the **thyroid gland**. It stimulates the **basal metabolic** including:
- the rate of respiration in cells
- how quickly substances are broken down
- how quickly new substances are synthesised.

Thyroxine also plays an important role in growth and development. For example, it regulates the growth of bones and the way in which the brain matures.

Thyroxine production is controlled by **TSH** (thyroid-stimulating hormone), produced in the **pituitary gland**.

Control of thyroxine

The control of thyroxine levels by TSH is an example of **negative feedback**.

pituitary

If thyroxine levels are low, TSH levels rise, so thyroxine levels increase.

If thyroxine levels are high, TSH levels fall, so thyroxine levels decrease.

thyroid

Worked example

Explain how the production of thyroxine is controlled by negative feedback. **(4 marks)**

When the concentration of thyroxine in the blood is too low, more TSH is released from the pituitary gland. This causes the thyroid gland to produce more thyroxine.

On the other hand, if the concentration of thyroxine in the blood is too high, less TSH is released from the pituitary gland, so the thyroid produces less thyroxine.

Note that the level of adrenaline is **not** controlled by negative feedback. The adrenal glands stop releasing adrenaline when danger is past, reducing levels to normal again.

Negative feedback

Negative feedback mechanisms have these features in common:
- A change in the level of a factor causes responses that reverse the change.
- The level of the factor is maintained within a small range.

Factor rises above normal. → corrective mechanism + → Factor falls towards normal.

factor at normal level − → corrective mechanism

factor at normal level

Factor falls below normal. → corrective mechanism + → Factor rises towards normal.

Now try this

Some people have an overactive thyroid gland. The symptoms of this include weight loss despite eating more, and a swollen neck. Suggest reasons for these symptoms. **(4 marks)**

Plant hormones

Plants produce hormones to control and coordinate their growth and development.

Tropisms

Plants respond to changes in light, moisture and gravity by changing how they grow:
- Growth in response to light is called **phototropism**.
- Growth in response to gravity is called **gravitropism** or **geotropism**.

Growth towards a stimulus is a **positive tropism**, and growth away from a stimulus is a **negative tropism**. Plant hormones called **auxins** control these tropisms.

Shoots grow:
- towards light
- against gravity.

Roots grow:
- away from the light
- in the direction of gravity.

Tropisms in roots

Auxins **inhibit** cell elongation in roots.

1 Roots show **negative phototropism**:
- Auxins move towards the shaded part of the root.
- Lit side becomes longer.
- Root grows away from the light.

2 Roots show **positive gravitropism**:
- Auxins move towards the lower part of the root.
- Upper side becomes longer.
- Root grows downwards.

Roots and shoots grow straight if the auxin levels are equal on both sides.

Tropisms in shoots

Auxins **stimulate** cell elongation in shoots.

1 Shoots show **positive phototropism**:
- Auxins move towards the shaded part of the shoot.
- Shaded side becomes longer.
- Shoot grows towards the light.

2 Shoots show **negative gravitropism**:
- Auxins move towards the lower part of the shoot.
- Lower side becomes longer.
- Shoot grows upwards.

Notice that opposite processes happen in roots and shoots.

Worked example

This shoot is growing towards the light. Explain what is happening to auxin at point A, and to the cells at point B. **(3 marks)**

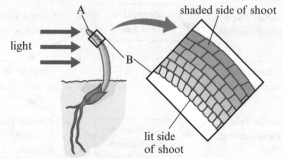

light

A

B

shaded side of shoot

lit side of shoot

At point A, light is causing the auxin to move to the shaded side of the shoot.

At point B, cells on the shaded side contain more auxin than cells on the lit side. So the cells of the shaded side grow longer and the shoot grows towards the light.

Light is needed for **photosynthesis**. This response enables the shoot to grow towards the light, so that photosynthesis can happen at a greater rate.

Now try this

 1 Describe the difference in the response of root cells and shoot cells to auxins. **(2 marks)**

 2 Explain the advantages of tropisms to plant roots. **(4 marks)**

Investigating plant responses

 Practical skills You can investigate the effect of **light** or **gravity** on the growth of newly germinated seedlings.

Core practical

Investigating the effect of light intensity on the growth of seedlings

Aim

to investigate the effect of light on the height of seedlings

Apparatus

- cress seeds
- Petri dishes
- cotton wool
- ruler, marked in 1 mm scale divisions
- water
- bright windowsill and dark cupboard

Method

1 Spread some cotton wool on the bottom of three Petri dishes, and wet it with water.
2 Add about 10 seeds to each dish, then put the dishes in a warm place to allow the seeds to germinate.
3 Once the seeds have germinated, remove any excess seedlings so each dish contains the same number of seedlings.
4 Place one dish on the windowsill, one in a cupboard and one in a dim place.
5 Every day, measure the height of each seedling. Continue for at least one week.

Results

Record your results in three suitable tables, showing the height of each seedling each day.

Analysis

1 Calculate the daily mean heights of the seedlings in each Petri dish.
2 Plot a graph with:
   - mean height on the vertical axis
   - day on the horizontal axis.
   Plot the results for all three Petri dishes on the same axes, and label the conditions.
3 Describe what the results show about the effect of light on the height of seedlings.

White mustard seeds could be used instead.

Dim light conditions could be achieved by placing a Petri dish on the windowsill or in the cupboard on alternate days.

You may need to add more water from time to time to stop the cotton wool drying out.

Rest the end of the ruler on the cotton wool and hold it gently against the seedling.

The effects of gravity

You could also investigate the effects of gravity on growth:
- Roll up a paper towel, place it in a beaker, and dampen it with water.
- Place a germinated bean seed between the paper and the glass.
- Allow it to grow so the root and shoot are visible, then lie the beaker on its side.

Observe and record (with labelled biological drawings) what happens to the root and shoot over the next few days.

Now try this

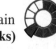 1 Describe why each Petri dish must contain several seedlings. **(2 marks)**

2 The scale on most rulers begins a few millimetres from the end. Explain the effect of this on the accuracy of the height measurements. **(3 marks)**

Uses of plant hormones

Plant hormones are used in agriculture, horticulture and the food industry.

Uses of auxins

Selective **weedkillers** contain **auxins**.

Weed plants compete with crop plants for water and minerals from the soil.

narrow-leaved crop plant

broad-leaved weed plant

Broad-leaved plants absorb more of the auxins than narrow-leaved plants. This causes them to grow too quickly and die.

Auxins stimulate the elongation of cells in shoots, and inhibit this process in roots.

The crop plants get more water and minerals, and so grow better.

Plants can be **cloned** by gardeners and scientists:

- Gardeners take **cuttings** (usually a shoot with leaves), which grow into new plants.
- Scientists use **tissue culture** to produce new plants from tiny pieces of a plant.

Gardeners use **rooting powder**, which contains auxins, to help roots develop and the cuttings grow into new plants.

Plant scientists use auxins to promote the growth of the plantlets during tissue culture.

You can revise details of these methods to produce genetically identical plants on page 88.

Worked example

Tomatoes are often picked while they are green and unripe. They are then transported to the distributors where they are exposed to ethene gas.

(a) Suggest a reason why the tomatoes are transported before they are ripe. **(2 marks)**

The fruit is still quite hard at this stage, so it is less likely to be damaged during handling and transport.

(b) Explain why the tomatoes are exposed to ethene before transport to the shops.

(2 marks)

Ethene controls ripening of fruits. It causes the green unripe tomatoes to ripen and turn red, ready for sale.

Gibberellins

Gibberellins are substances that affect different processes in plants. They are important in starting the germination of seeds. Gibberellins are used in the beer industry to speed up the germination of barley seeds to make malt, an ingredient of beer.

Gibberellins are also used to:

- produce large grapes and other fruit
- start flowering in plants, for sale in garden centres and flower shops.

Ethene also controls cell division in fruits. It is produced naturally by ripening fruit, including bananas and peaches.

Now try this

1 State **two** uses of auxins in agriculture and horticulture. **(2 marks)**

2 Explain why fruit ripens more quickly when it is in a bowl with bananas. **(2 marks)**

Agriculture is to do with farming, and horticulture is to do with gardening.

Extended response – Homeostasis and response

There will be at least one 6-mark question on your exam paper. For these questions, you will need to think scientifically and structure your answer logically, showing how the points you make are related to each other.

You can revise the topic for this question, which is about plant hormones, on page 72.

Worked example

A student sows a seed in a pot of soil. The diagram shows what happens after the seed germinates. The shoot grows upwards and bends towards the light from a window. The roots grow downwards into the soil. Compare the responses that cause the plant to grow in this way. In your answer, name the responses and describe how they happen.

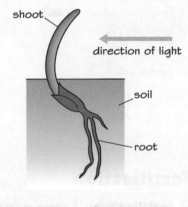

shoot

direction of light

soil

root

Command word: Compare

You need to describe the similarities and/or differences between things, not just write about one thing.

The response of a plant to gravity is called gravitropism. Roots show positive gravitropism because they grow downwards. On the other hand, shoots show negative gravitropism because they grow upwards.

The response of a plant to light is called phototropism. Shoots show positive phototropism because they grow towards the light, and roots show negative phototropism.

Tropisms happen because of substances called auxins. These are plant hormones and they affect the way plant cells grow. High auxin concentrations stop root cells growing longer, but they stimulate shoot cells to grow longer.

The auxin concentration is greater on the bottom side of a root, so its top side grows longer and the root bends downwards. The auxin concentration is greater on the shaded side of a shoot, so this side grows longer and the shoot bends towards the light.

 The word 'geotropism' could be used instead of 'gravitropism'.

The answer gives a difference between a positive and negative tropism.

Here the answer gives a similarity and a difference between roots and shoots.

This part of the answer includes an explanation. Instead it could have just compared where, in roots and shoots, high auxin concentration and cell elongation occur.

You need to organise your answer into a logical order, with ideas linked in a sensible way. In this answer, responses are named as required by the question, and several similarities and differences are given.

Think about the route, including the three types of neurones involved and the effector. You can revise this topic on page 55.

Now try this

When a bright light shines into the eye, muscles in the iris contract. The size of the pupil decreases so that less light can enter the eye. This is an example of a reflex.
Describe how impulses travel through the reflex arc involved, starting with receptor cells in the eye. **(6 marks)**

Meiosis

Cells in reproductive organs divide by **meiosis** to form **gametes**.

Meiosis outlined

When a cell divides by meiosis, the four cells produced are all genetically different.

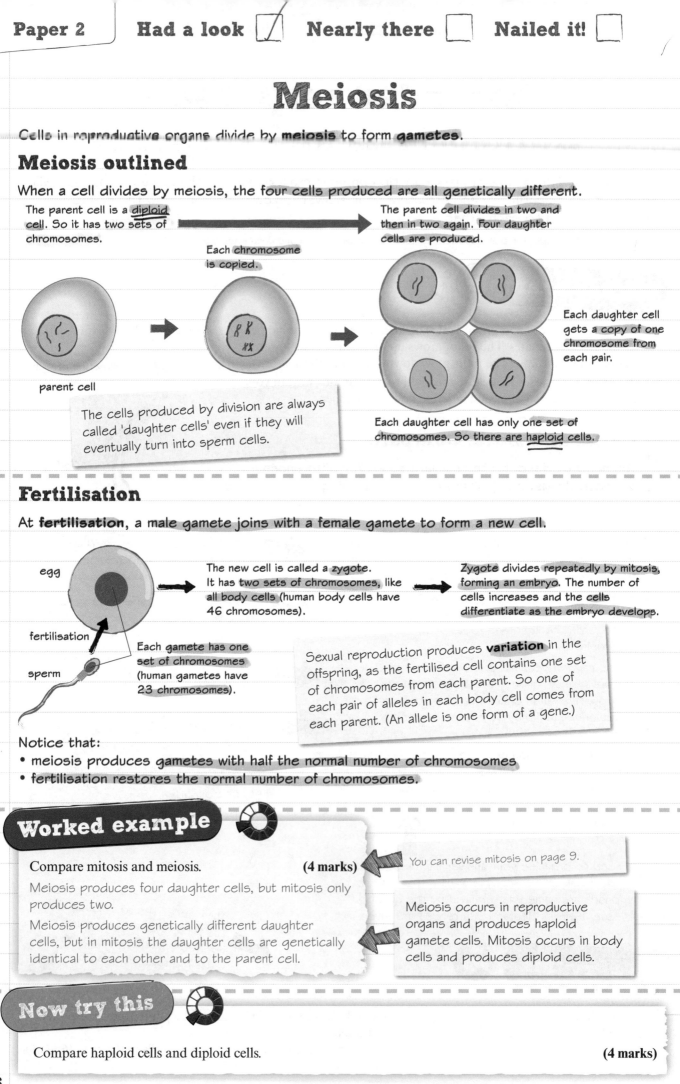

The parent cell is a diploid cell. So it has two sets of chromosomes.

Each chromosome is copied.

The parent cell divides in two and then in two again. Four daughter cells are produced.

Each daughter cell gets a copy of one chromosome from each pair.

parent cell

The cells produced by division are always called 'daughter cells' even if they will eventually turn into sperm cells.

Each daughter cell has only one set of chromosomes. So there are haploid cells.

Fertilisation

At **fertilisation**, a male gamete joins with a female gamete to form a new cell.

egg

fertilisation

sperm

Each gamete has one set of chromosomes (human gametes have 23 chromosomes).

The new cell is called a zygote. It has two sets of chromosomes, like all body cells (human body cells have 46 chromosomes).

Zygote divides repeatedly by mitosis, forming an embryo. The number of cells increases and the cells differentiate as the embryo develops.

Sexual reproduction produces **variation** in the offspring, as the fertilised cell contains one set of chromosomes from each parent. So one of each pair of alleles in each body cell comes from each parent. (An allele is one form of a gene.)

Notice that:
- meiosis produces gametes with half the normal number of chromosomes
- fertilisation restores the normal number of chromosomes.

Worked example

Compare mitosis and meiosis. **(4 marks)**

You can revise mitosis on page 9.

Meiosis produces four daughter cells, but mitosis only produces two.

Meiosis produces genetically different daughter cells, but in mitosis the daughter cells are genetically identical to each other and to the parent cell.

Meiosis occurs in reproductive organs and produces haploid gamete cells. Mitosis occurs in body cells and produces diploid cells.

Now try this

Compare haploid cells and diploid cells. **(4 marks)**

Sexual and asexual reproduction

New individuals can form by sexual reproduction or by asexual reproduction.

Features of reproduction

	Sexual	Asexual
Number of parents	two	one
Meiosis occurs?	✓	✗
Gametes fuse?	✓	✗
Mitosis occurs?	✓	✓
Genetic information	mixing occurs	no mixing
Offspring	not identical	identical

Gametes form by meiosis in sexual reproduction:
- Male gametes are **sperm** cells in animals and **pollen** cells in flowering plants.
- Female gametes are **egg** cells in animals and flowering plants.

Mitosis forms **clones**, genetically identical offspring, in asexual reproduction. It increases the number of cells in both types of reproduction.

Sexual reproduction

Advantages of sexual reproduction include:
- 👍 It produces **variation** in offspring. adaptation
- 👍 If the environment changes, variation gives the species a survival advantage by natural selection (see page 84).
- 👍 Humans can increase food production by using selective breeding to speed up natural selection (see page 85). increase food production

Disadvantages include:
- 👎 Individuals must find a mate, which takes time and energy.

Asexual reproduction

Advantages of asexual reproduction include:
- 👍 Individuals do not need a mate, which is efficient in terms of time and energy. safe time
- 👍 It is faster than sexual reproduction, so many offspring can be produced quickly when the environment is favourable. faster

Disadvantages include:
- 👎 The parent and its offspring are all genetically identical, so if the environment changes they may be poorly adapted to the new conditions and all may die.

Worked example

Many plants produce seeds by sexual reproduction, but also reproduce asexually. For example, strawberry plants produce runners with small plants along them. Daffodils produce underground bulbs which divide to form new bulbs. Suggest the advantages of reproducing like this. **(4 marks)**

A strawberry plant that is adapted to its environment can quickly spread by asexual reproduction. Daffodils can form new plants, even if some are damaged by frost and fail to be pollinated so they can form seeds. The use of sexual reproduction introduces variation in the plant population, making it more likely to survive changes in conditions.

Fungi and *Plasmodium*

Many fungi reproduce asexually by producing spores in their familiar mushrooms and toadstools. They also reproduce sexually to give variation in the population.

Plasmodium, the parasite that causes malaria, reproduces asexually in its human host but sexually in the mosquito vector.

Plants usually have adaptations that cause their seeds to fall far from the parent plant. This helps to spread the species. It also reduces competition between the parent and its offspring for light, space, water and other resources.

Now try this

1   Describe **two** ways in which plants naturally produce clones. **(3 marks)**

2   Explain how reproduction can introduce variation into a population. **(4 marks)**

DNA and the genome

DNA is genetic material. An organism's **genome** is its entire genetic material.

DNA structure

The nucleus contains chromosomes.

Most cells have a nucleus.

eukaryotic cell

chromosome

DNA is the genetic material. It is tightly coiled in a chromosome.

A **gene** is a small section of DNA on a chromosome. Each gene codes for a particular sequence of amino acids to make a specific protein.

DNA

polymer made up of 2 strands forming double helix structure

DNA is a **polymer** made up of two strands, forming a double helix structure.

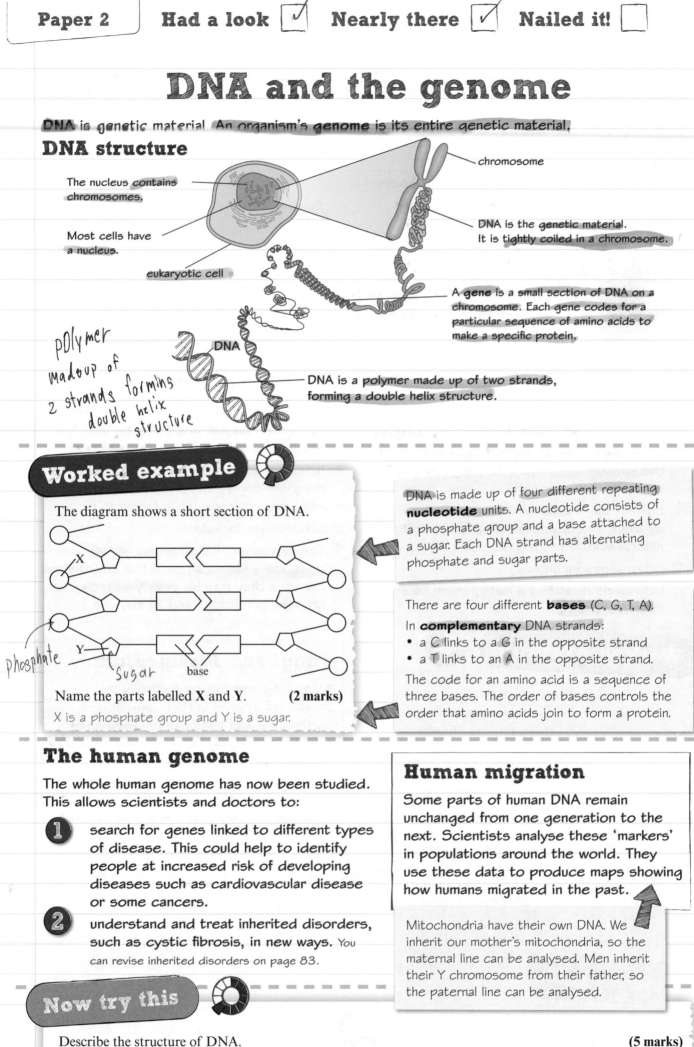

Worked example

The diagram shows a short section of DNA.

Phosphate

X

Y

Sugar   base

Name the parts labelled **X** and **Y**.   **(2 marks)**

X is a phosphate group and Y is a sugar.

DNA is made up of four different repeating **nucleotide** units. A nucleotide consists of a phosphate group and a base attached to a sugar. Each DNA strand has alternating phosphate and sugar parts.

There are four different **bases** (C, G, T, A).

In **complementary** DNA strands:
• a C links to a G in the opposite strand
• a T links to an A in the opposite strand.

The code for an amino acid is a sequence of three bases. The order of bases controls the order that amino acids join to form a protein.

The human genome

The whole human genome has now been studied. This allows scientists and doctors to:

1 search for genes linked to different types of disease. This could help to identify people at increased risk of developing diseases such as cardiovascular disease or some cancers.

2 understand and treat inherited disorders, such as cystic fibrosis, in new ways. You can revise inherited disorders on page 83.

Human migration

Some parts of human DNA remain unchanged from one generation to the next. Scientists analyse these 'markers' in populations around the world. They use these data to produce maps showing how humans migrated in the past.

Mitochondria have their own DNA. We inherit our mother's mitochondria, so the maternal line can be analysed. Men inherit their Y chromosome from their father, so the paternal line can be analysed.

Now try this

Describe the structure of DNA.   **(5 marks)**

Protein synthesis

Making a protein

The **sequence** or order of the **bases** in DNA is used to produce a template molecule.
This leaves the nucleus so **protein synthesis** can begin (step 1 in the diagram).

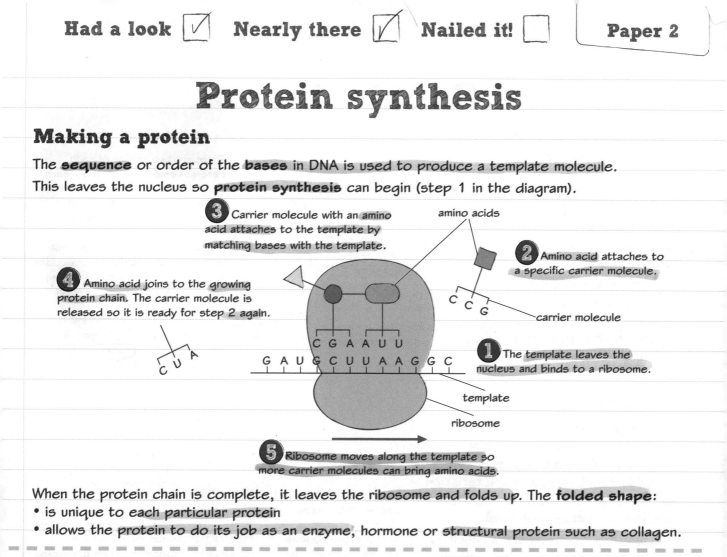

3 Carrier molecule with an amino acid attaches to the template by matching bases with the template.

amino acids

2 Amino acid attaches to a specific carrier molecule.

carrier molecule

4 Amino acid joins to the growing protein chain. The carrier molecule is released so it is ready for step 2 again.

C C G

C U A

C G A A U U

G A U G C U U A A G G C

1 The template leaves the nucleus and binds to a ribosome.

template

ribosome

5 Ribosome moves along the template so more carrier molecules can bring amino acids.

When the protein chain is complete, it leaves the ribosome and folds up. The **folded shape**:
- is unique to each particular protein
- allows the protein to do its job as an enzyme, hormone or structural protein such as collagen.

Effects of mutations

Mutation changes the sequence of DNA bases.

1 The change may have no effect on the protein made. The order of amino acids may be the same, or be so slightly different that the function of the protein is unchanged.

2 The change may change the order of amino acids in such a way that the shape of the protein is changed. This may affect the active site of an enzyme so it no longer works, or it may weaken a structural protein.

Mutations

– sub change
–

Mutations can include changes such as:
- deletion of a part of the DNA
- repeating a part of the DNA
- a small change in the sequence of bases.

Mutations happen continuously. They can happen when the DNA is copied before cell division. They can also happen because of **mutagens**, which include:
- substances that change DNA
- X-rays and ultraviolet light.

 Worked example

Not all parts of DNA code for proteins. Explain the effects of variations in this DNA. **(2 marks)**

These non-coding parts can switch genes on and off. Variations in these parts may affect how and when genes are expressed.

The human genome has about 20 000 genes, but most of it consists of non-coding DNA.

By controlling the expression of genes, the non-coding DNA may also affect an organism's **phenotype** – its appearance and functions.

 Now try this

Describe how protein synthesis happens. **(6 marks)**

Genetic terms

You need to be able to explain all the genetic terms shown in bold on this page.

Inside a cell

A **gamete** is a sex cell formed by meiosis in a reproductive organ.

At fertilisation:

- sperm cells and egg cells fuse (join together) in animals
- pollen cells and egg cells fuse in flowering plants.

cell

Most cells contain a nucleus in which there are **chromosomes**.

There are two copies of each **chromosome** in body cells – each copy has the same genes in the same order along its length (except chromosomes that determine sex).

A **gene** is a small section of DNA on a chromosome. Each gene codes for a particular sequence of amino acids to make a specific protein.

A gene may come in different forms, called **alleles**, that produce different variations of the characteristic, e.g. different eye colours.

Alleles

The **genotype** is the **alleles** of a particular gene present in an individual. The **phenotype** is an individual's observed characteristics or traits, produced by these alleles working at a molecular level.

Most characteristics are a result of multiple **genes** interacting, rather than the result of a single gene.

Here there are different **alleles** of the same gene, so the organism will be **heterozygous** for the trait controlled by the gene.

Chromosomes of the same pair have the same genes in the same order.

These two genes have the same alleles, so the organism will be **homozygous** for the traits they control.

Worked example

Fur colour in mice is controlled by a single gene. This has two alleles, shown as B and b. The table shows the three possible genotypes and phenotypes for this trait.

Genotype	Phenotype
BB	black fur
Bb	black fur
bb	brown fur

(a) Explain why the allele for black fur is described as a **dominant** allele. **(2 marks)**

Mice have black fur whether they have two copies of this allele, BB, or just one, Bb.

(b) Explain why the allele for brown fur is described as a **recessive** allele. **(2 marks)**

Mice only have brown fur when they have two copies of this allele, bb. They have black fur instead if they only have one copy, Bb.

Red–green colour blindness in humans is also controlled by a single **gene**.

A **dominant allele** is always expressed, even if only one copy is present:
- BB mice are homozygous for the B allele.
- Bb mice are **heterozygous**.

A **recessive allele** is only expressed if two copies are present (so no dominant allele):
- bb mice are **homozygous** for the b allele.

Now try this

Flower colour in pea plants is controlled by two alleles, R and r. RR and Rr plants have purple flowers and rr plants have white flowers. Explain which allele is dominant and which is recessive. **(4 marks)**

Genetic crosses

The results of **genetic crosses** are predicted using genetic diagrams and Punnett square diagrams.

Genetic diagrams

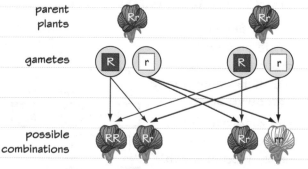

parent plants

gametes

possible combinations

In the **genetic diagram** on the right, both parent plants are heterozygous for flower colour. It shows that the possible combinations of alleles for this Rr × Rr genetic cross are:

• RR, purple flowers
• Rr, purple flowers (two ways to get this)
• rr, white flowers.

Probability

Notice in the two diagrams on the right that:
• ¼ combinations are RR (purple flowers)
• ¼ combinations are rr (white flowers).

There are two combinations that give Rr, so:
• ²⁄₄ combinations are Rr (purple flowers).

The **probability** that these parents will produce offspring with purple flowers is:

(¼ + ²⁄₄) = ¾

You could also write this as 0.75, 75%, 3 in 4 or 3:1. The actual outcome may differ, particularly if few offspring are produced.

Punnett square diagrams

You can also show the outcomes of the Rr × Rr genetic cross above using a **Punnett square diagram**.

		Parent gametes	
		R	r
Parent gametes	R	RR	Rr
	r	Rr	rr

You should be able to complete or draw diagrams like these.

Worked example

The allele for green seed pods, G, is dominant to the allele for yellow seed pods, g.

(a) Draw a Punnett square diagram to show the possible outcomes of a Gg × gg genetic cross. **(2 marks)**

Gametes	G	g
g	Gg	gg
g	Gg	gg

It may help the answer to part (b) if the seed pod colour is written underneath each outcome.

(b) Identify the offspring that will produce yellow seed pods, and explain your answer. **(2 marks)**

The gg offspring will have yellow seed pods. This is because the allele for yellow, g, is recessive. It will only be expressed if an individual has two copies.

The probability of having an offspring that produces yellow seed pods is ²⁄₄. You could also write this as 0.5, 50%, 1 in 2 or 1:1.

Now try this

The allele for black fur in rabbits, B, is dominant. The allele for brown fur, b, is recessive.

(a) Draw a Punnett square diagram to show the possible outcomes of a BB × bb genetic cross. **(2 marks)**

(b) State the probability that this cross will produce baby rabbits with brown fur, and explain your answer. **(3 marks)**

Use the outcomes in your Punnett square diagram to help you.

Family trees

Family trees can be analysed to study the inheritance of alleles.

Analysing a family tree

Cystic fibrosis, CF, is an inherited disorder (see page 83). It is caused by a recessive allele.

The diagram shows an example of a family tree in which some members have cystic fibrosis.

Key

- ☐ male
- ■ male with cystic fibrosis
- ○ female
- ● female with cystic fibrosis

Ethan, one of Arun and Beth's three children, has cystic fibrosis:

- The allele for cystic fibrosis is recessive, so Ethan must have inherited one copy from each parent.
- Arun and Beth must be heterozygous for the cystic fibrosis allele.

Mia, one of Gill and Harry's two daughters, has cystic fibrosis:

- The allele for cystic fibrosis is recessive, so Mia must have inherited one copy from each parent.
- Gill and Harry must be heterozygous for the cystic fibrosis allele.

Worked example

The ability to taste phenylthiocarbamide (PTC) is a dominant condition. The diagram shows the inheritance of PTC tasting in a family. Describe the evidence that PTC tasting is controlled by a dominant allele.
(2 marks)

non-PTC tasters   PTC tasters

Individuals 1 and 2 are PTC tasters, but they have two children (individuals 6 and 8) who are non-tasters. Therefore, PTC tasting must be controlled by a dominant allele and non-tasting by a recessive allele.

Trees and probability

You can use a **Punnett square diagram** (page 81) to calculate the probability that Gill and Harry will have a child with cystic fibrosis.

Gametes	F	f
F	FF	Ff
f	Ff	ff

A child with cystic fibrosis will be ff, as the allele is recessive so two copies must be present.

There is one ff combination from the four possible combinations, so the **probability** of having a child with cystic fibrosis is:

¼ (0.25, 25%, 1 in 4 or 1:3).

In this family, the actual outcome is ½ (0.5, 50%, 1 in 2 or 1:1).

Now try this

The question is about the family tree in the worked example.
(a) Give the genotypes for individuals 9 and 10, and explain your answer. **(4 marks)**

Use T for the dominant allele and t for the recessive allele.

(b) Use a Punnet square diagram to calculate the probability that the people in part (a) will have a child who can taste PTC. **(4 marks)**

You can revise these diagrams on page 81.

Inheritance

Some disorders and the sex of an individual are inherited.

Polydactyly

Polydactyly is an **inherited disorder** in which people have extra fingers or toes. It is caused by a dominant allele. This means that:
* people have polydactyly if they inherit one **or** two copies of the allele.

The Punnett square diagrams show the two situations where one parent has polydactyly.

> Polydactyly does not usually harm health. Other dominant genetic disorders may lead to ill-health, such as Huntington's disease.

If both parents are homozygous (one with polydactyly and one without):
* all the children will be heterozygous so will have polydactyly.

Gametes	P	P
p	Pp	Pp
p	Pp	Pp

If one parent is heterozygous (one with polydactyly and one without):
* the probability of having a child with poly-dactyly is ½ (0.5, 50%, 1 in 2 or 1:1).

Gametes	P	p
p	Pp	pp
p	Pp	pp

Cystic fibrosis

Cystic fibrosis is an inherited disorder of the cell membranes. It is caused by a recessive allele. This means that:
* People have cystic fibrosis only if they inherit two copies of the allele.
* People who are heterozygous for the trait are carriers – they can pass the allele to their children but do not have the disorder themselves.

If both parents are heterozygous (carriers):
* the probability of having a child with cystic fibrosis is ¼ (0.25, 25%, 1 in 4 or 1:3).

Gametes	F	f
F	FF	Ff
f	Ff	ff

carrier (like the parents)   child with cystic fibrosis

Worked example

Draw a Punnett square diagram to show how sex in humans is inherited. Identify the sex of each offspring in your diagram. **(3 marks)**

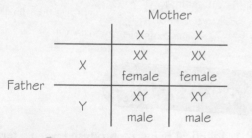

	Mother	
	X	X
Father X	XX female	XX female
Y	XY male	XY male

> Ordinary human body cells have 23 pairs of chromosomes. One of these pairs carries the genes that determine sex:
> * females – same sex chromosomes, XX
> * males – different sex chromosomes, XY.

> The probability of a couple having a girl is ½ or 50%. This is the same as the probability of having a boy.

Now try this

1. Explain why all human egg cells contain the same sex chromosome. **(2 marks)**

2. Describe **two** differences between the inheritance of polydactyly and the inheritance of cystic fibrosis. **(2 marks)**

Variation and evolution

Evolution is a change in the inherited characteristics of a population over time due to natural selection. It may result in the formation of a new species.

Causes of variation

A **population** is all the individuals of a species living in the same place at the same time.

The differences in the characteristics of the individuals in a population are called **variation**.

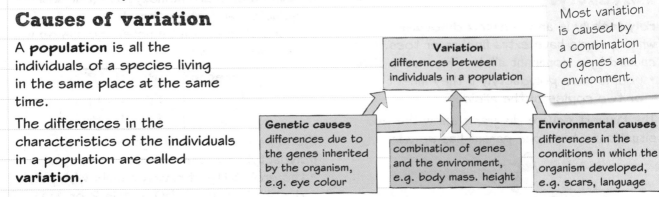

Most variation is caused by a combination of genes and environment.

Variation
differences between individuals in a population

Genetic causes
differences due to the genes inherited by the organism, e.g. eye colour

combination of genes and the environment, e.g. body mass, height

Environmental causes
differences in the conditions in which the organism developed, e.g. scars, language

Evolution

The theory of evolution by **natural selection** states that all species of living things have evolved from simple life forms, which first developed over three billion years ago.

Individuals in a population show variation in their inherited characteristics.

⬇

Some individuals have phenotypes that make them better suited to their environment.

⬇

These individuals are more likely to survive and reproduce, so pass their alleles to their offspring.

⬇

These offspring are more likely to have phenotypes that make them better suited to the environment. They are more likely to survive and reproduce.

Variation and mutation

There is usually extensive **genetic variation** within a population of a species.

All genetic **variants** arise from **mutations**. Of these mutations:
- most have no effect on the **phenotype**
- some influence the phenotype
- very few determine the phenotype.

Mutations occur continuously, but very rarely a mutation may lead to a new phenotype. If this is suited to an environmental change, it can lead to a relatively rapid change in a species.

◀ Two populations of one species may become too different to breed with each other any more to produce fertile offspring. They will have formed two species.

This is an example of where you may need to apply your knowledge and understanding to an unfamiliar example or situation.

Poachers hunt African elephants illegally for their tusks. A mutation may cause elephants to be born tuskless. In Uganda in 1930, only 1% of elephants were born tuskless. By 2010, 15% of female elephants and 9% of male elephants were born without tusks. Explain this change using the theory of evolution. **(3 marks)**

Elephants with tusks are more likely to be poached and killed, so do not survive to pass on their alleles. Tuskless elephants are more likely to survive to pass on the allele for no tusks, so the allele for no tusks increases in frequency in the population.

Now try this

1 Identical twins are produced when a fertilised egg cell divides in two. Explain why identical twins share many, but not all, of their characteristics. **(3 marks)**

2 In a study, 82% of head lice were found to be resistant to substances used to treat head lice infestations. Use your knowledge of natural selection to explain how this has happened. **(4 marks)**

Selective breeding

Humans have been using **selective breeding** of plants and animals for thousands of years.

The process

Selective breeding is also called **artificial selection**. Humans use this process to breed plants and animals that have particular genetic characteristics. The process of selective breeding takes place over many generations, not just one.

Select individuals with the desired characteristic from a mixed population.

↓

Breed from the selected individuals.

↓

Select offspring with the desired charateristics.

→

Repeat over many generations until all the offspring show the desired characteristic.

Worked example

Describe how a farmer would produce a herd of cattle that has a high milk yield. **(4 marks)**

The farmer would choose a cow that produced a lot of milk. This cow would then be bred with a bull whose mother also had a high milk yield. Next, the farmer would choose the female offspring that have a high milk yield, and breed them with a bull whose mother also had a high milk yield. The farmer would do this for many generations.

Characteristics can be chosen for usefulness or appearance, including:
- animals that produce more milk or meat
- food crops that resist plant diseases
- domestic dogs that have a gentle nature
- plants with large or unusual flowers.

Worked example

The graph shows the mean milk yield per cow from 1957 to 2007. The change is mainly the result of selective breeding. Over the same period, there was an increase in the percentage of cows with mastitis (sore udders) and leg problems. This was reported in a farming newspaper in an article with the title 'Selective breeding has improved dairy farming'.

Evaluate the statement in the title. **(5 marks)**

The graph shows that the milk yield per cow has more than doubled between 1957 and 2007. This is useful because farmers can produce more milk from fewer cows, reducing costs.

However, more cows are having problems with their legs and udders. This may be because they are producing so much milk, so this is an animal welfare issue. It will also cost the farmer more in veterinary bills.

Overall, I think that selective breeding has improved dairy farming, but the animal welfare issues it has caused must be sorted out.

This is an advantage to the farmer.

This describes an **ethical issue** and a disadvantage to the farmer.

Your conclusion should relate to information given, and benefits and risks identified.

Inbreeding

Selective breeding, especially if it involves close relatives, can lead to **inbreeding**. Some breeds may be particularly likely to suffer from disease or inherited defects.

Describe how a breed of sheep that grow a lot of wool could be produced. **(4 marks)**

Genetic engineering

Genetic engineering is a process in which an organism's **genome** is modified to give a desired characteristic by introducing a gene from another organism.

The process

The flow chart outlines the process of genetic engineering.

| Genes can be transferred from any kind of organism to any other kind of organism, e.g. bacteria, humans, other animals, plants. | The gene for a characteristic is 'cut out' of a chromosome using enzymes. | → | The gene is transferred to a cell of another organism, and inserted into a chromosome. | → | The cell of this organism now produces the characteristic from the gene. |

GM crops

Genetically engineered crop plants are called **genetically modified** (GM) crops.

GM crops include plants that are:

👍 resistant to plant diseases

👍 resistant to attack by insect pests

👍 resistant to herbicides, so that fields can be sprayed to kill weeds, but not the crop plants.

These characteristics help to produce increased **yields**, including bigger and better fruits.

Concerns about GM crops include:

👎 the possible effects on populations of wild flowers due to herbicides

👎 the possible effects on populations of insects if there is less food for them

👎 whether the possible effects on human health of eating GM crops has been researched enough.

Worked example

Patients with diabetes may be treated with human insulin. Bacteria have been genetically engineered to carry the gene for human insulin. These bacteria are grown in large numbers to produce relatively large amounts of human insulin.

(a) Explain why the bacteria had to be modified to make human insulin. **(2 marks)**

Bacteria do not naturally have the gene to produce human insulin. Genetic engineering must be used to introduce the gene into their genome.

(b) Describe how the genetically engineered bacteria were produced. **(2 marks)**

The gene for making human insulin was cut out of a human chromosome. The gene was then inserted into bacterial cells so it would become part of their genome. The modified bacteria would then be able to make insulin.

Engineering sheep

Sheep have been genetically engineered to produce a human protein in their milk. Some people have an inherited disorder that stops them making this protein, causing lung and liver problems. The protein in sheep's milk can be used to treat them.

Scientists are researching ways to genetically modify cells in people with inherited disorders. They hope that this treatment would overcome the symptoms of these disorders. However, some people have objections to genetic engineering, including ethical concerns.

You can revise the details of the main steps in genetic engineering on page 87.

Now try this

1 Give the meaning of **genetic engineering**. **(3 marks)**

2 Describe how a GM crop with herbicide resistance could be developed. **(3 marks)**

Stages in genetic engineering

Genetic engineering involves **enzymes** to cut and join DNA, and **vectors** to insert modified DNA into cells.

The main stages

1 A gene (e.g. for insulin) is cut out of a human chromosome using enzymes.

insulin gene   ✂ cutting enzymes   ▱ sticking enzymes

human chromosome

3 The human insulin gene and the plasmid are mixed together.

plasmid

bacterium

4 The human insulin gene and the plasmid are stuck together to make a new plasmid.

2 A DNA plasmid is taken out of a bacterium and cut open using enzymes.

5 The new plasmid with the human insulin gene is put back into a bacterium. The bacterium has been **genetically modified**. The bacterium will now make human insulin.

Vectors

A **vector** in genetic engineering is an organism or part of an organism that can enter a cell and insert DNA into the cell's DNA. Vectors include:

- bacterial plasmids – circles of DNA that can move into and out of bacteria cells
- viruses that can enter plant or animal cells, carrying DNA with them.

Different vectors are used, depending on the type of cell to be modified and the size of the gene to be inserted.

Enzyme cuts out the gene for the desired characteristic.

GATC

CTAG

complementary sequence

Enzyme cuts open the vector DNA giving ends with the same sequence.

CTAG   GATC

Enzymes

Some **enzymes** can:

- bind to certain sequences of DNA **bases**
- cut or join a DNA strand.

The end of one DNA strand is **complementary** to the end of another DNA strand, so joins to it.

C links to G, and T links to A.

Gene and cut vector are mixed, and the two pieces of DNA join together via the complementary bases.

Worked example

Describe the main steps in genetic engineering. **(3 marks)**

Enzymes are used to isolate the gene for the desired characteristic. This gene is inserted into a vector, such as a bacterial plasmid or a virus. The vector is then used to insert the gene into the required cells.

Genes can be transferred to the cells of animals, plants or microorganisms.

This is done at an early stage in their development so that the genetically engineered organism develops with the desired characteristic.

Now try this

Give the meaning of a **vector**, as used in genetic engineering.

(2 marks)

Cloning

Plants and animals can be **cloned** to produce large numbers of genetically identical individuals.

Cloning plants

The diagrams show the stages involved in **tissue culture**. This is important for:
- preserving rare species of plants
- producing many plants in plant nurseries.

- Gardeners can produce many identical new plants by taking **cuttings** from a parent plant. You can revise the use of plant hormones in this older, simpler cloning technique on page 74.

tissue sample cut from parent plant → agar jelly containing plant hormones and nutrients → samples grow into tiny plants → trays of compost

Embryo transplant

This method of cloning animals involves:
- separating cells from a developing embryo at a stage before they become specialised
- transplanting these genetically identical embryos into host mothers.

Sperm are used to fertilise eggs.

Each fertilised egg develops into an embryo.

Embryo cells are separated before they become specialised.

Each embryo cell grows into a new embryo.

The offspring are clones of the original embryo.

Adult cell cloning

animal A   animal B

body cell, e.g. skin cell

unfertilised egg cell

rest of cell thrown away

nucleus removed and thrown away

nucleus removed

nucleus of body cell inserted into empty cell

cell with no nucleus

electric shock starts division

- Cell divides to form embryo.
- Embryo is inserted into womb of adult female C to develop until birth.

Worked example

The offspring inherits no components from its host mother, animal C.

The offspring inherits its mitochondria from animal B. Mitochondria have small amounts of their own DNA, so it is not fully genetically identical to animal A.

The diagram above shows how adult cell cloning works. The offspring is a clone of one of the three animals involved. Give the letter of this animal, and explain your answer. **(2 marks)**

Animal A. This is because its genes are contained in the nucleus taken from the body cell of animal A.

Now try this

Explain which method of animal cloning is likely to be used to produce many clones. **(2 marks)**

Darwin and Lamarck

The theory of evolution by natural selection developed over time using many scientists' work.

Charles Darwin (1809–1882)

Charles Darwin proposed the theory of **evolution** by **natural selection**. He published his ideas in his book *On the Origin of Species* in 1859. The main points of the theory are:

1 Individual organisms within a species show a wide range of variation for a characteristic.

2 Individuals with characteristics most suited to the environment are more likely to survive to breed successfully.

3 Characteristics that have enabled these individuals to survive are passed on to the next generation.

A timeline of events

1831–1836
The voyage of *HMS Beagle*. Darwin studies animals on the Galapagos Islands, including various finches.

⬇

1837
Darwin presents his collection of specimens to the Zoological Society.

⬇

1859
On the Origin of Species is published.

⬇

1871
The Descent of Man is published.

About Darwin's ideas

There was much controversy about Darwin's ideas, which were seen as revolutionary.

The theory of evolution by natural selection was only slowly accepted. Criticisms included:

☞ It challenged the idea that God made all the living things on Earth.

☞ At the time, there was not enough evidence to convince many scientists.

☞ At the time, the mechanism of inheritance and variation was not known (this was known 50 years later).

Although his theory was criticised, it was a result of scientific research, including:

👍 observations made on the *HMS Beagle* expedition around the world

👍 years of experiments and discussion with other scientists

👍 making use of developing scientific knowledge of fossils and geology.

Worked example

Jean-Baptiste Lamarck (1744–1829) proposed a theory of evolution by the inheritance of acquired characteristics:
- Body parts that are used become larger and stronger, while unused parts become smaller or disappear.
- Changes to an organism during its lifetime are passed to its offspring.

Use the information to suggest how Lamarck might have explained the evolution of long necks in giraffes. **(3 marks)**

A giraffe stretches upwards to reach leaves in high branches. The giraffe's neck becomes longer because it is used a lot. The giraffe's offspring inherit its long neck. Over time, giraffes develop increasingly long necks.

Alternative theories, like this one, are based mainly on the idea that changes to an organism during its lifetime can be inherited. We now know that this cannot happen in the vast majority of cases.

Now try this

Explain, using your knowledge and understanding of Darwin's theory of evolution by natural selection, how long necks evolved in giraffes. **(3 marks)**

Speciation

Speciation is the formation of new species.

Forming new species

Populations of the same species become separated e.g.
• they move to different islands
• a change in the environment produces a barrier to movement.

Other examples of **geographical isolation** include:
• new rivers forming
• volcanic eruptions
• mountain-building processes.

Natural selection in the different areas favours different variations of characteristics.

The original population will show variation between individuals due to genetic causes.

The characteristics of each population change over time.

Natural selection acting in the different environments produces gradual changes in each population.

Their characteristics become so different that the populations become different species.

Individuals from two different species are usually unable to produce fertile offspring.

Alfred Wallace (1823–1913)

Alfred Russel Wallace proposed the theory of evolution by natural selection:
• Like Charles Darwin, Wallace worked worldwide to gather evidence
• Wallace developed his theory independently of Darwin.

Later, in 1858, he published work jointly with Darwin. This prompted Darwin to publish his book *On the Origin of Species* a year later (you can revise Darwin's work on page 89).

Other work by Wallace

Wallace is best known for:
• his work on **warning colours** in animals such as caterpillars that are toxic to predators
• his theory of speciation.

Wallace proposed that, as two populations gradually differed, any **hybrids** would be less well adapted than their parents. This 'Wallace effect' would continue to drive speciation.

Scientists have gathered more evidence since, leading to our current scientific understanding.

Worked example

Anole lizards live on islands in the Caribbean. Each island has its own anole species. Islands that are further from the other islands have more species that are not found on other islands. DNA evidence suggests that many of the species are very closely related. Describe one way in which these species may have evolved. **(4 marks)**

A few lizards of one species moved from one island to another. The environment was different on this island, so natural selection there favoured different variations. Over time, the characteristics of the population on the new island changed enough so the lizards became a new species.

Many species of flightless birds live on the islands of New Zealand. All these species are related to species found in nearby Australia that can still fly. Suggest how the flightless bird species may have evolved. **(4 marks)**

There are no natural predators in New Zealand.

Mendel

Gregor Mendel (1822–1884) was the first person to discover the basics of **inheritance**.

Mendel's work

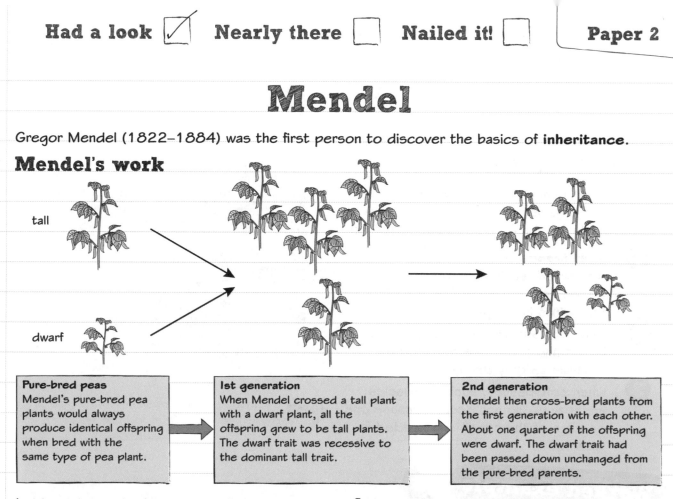

tall

dwarf

Pure-bred peas
Mendel's pure-bred pea plants would always produce identical offspring when bred with the same type of pea plant.

1st generation
When Mendel crossed a tall plant with a dwarf plant, all the offspring grew to be tall plants. The dwarf trait was recessive to the dominant tall trait.

2nd generation
Mendel then cross-bred plants from the first generation with each other. About one quarter of the offspring were dwarf. The dwarf trait had been passed down unchanged from the pure-bred parents.

👍 Mendel was the first person to show that, in the second generation, there were about three times more plants with the dominant characteristic than plants with the recessive characteristic.

👎 Few people read Mendel's work, published in 1865.

👎 It was only many years later that scientists linked Mendel's 'units' (inherited characteristics) with genes.

A timeline of discovery

The gene theory was developed from the work of many scientists over a long time.

Mid-19th century	**Late 19th century**	**Early 20th century**	**Mid-20th century**
Mendel shows that the inheritance of a characteristic is determined by 'units' that pass on unchanged.	The behaviour of chromosomes during cell division is observed.	'Units' and chromosomes are seen to behave in similar ways, so 'units' are genes located on chromosomes.	The structure of DNA is determined, and the mechanism of gene function is worked out.

Worked example

Mendel concluded from his experiments with pea plants that characteristics are caused by 'units' – separately inherited factors that were passed unchanged to descendants.
Explain Mendel's conclusion.   **(2 marks)**

Mendel showed that some characteristics of parent plants, such as dwarf size, seem to disappear in the first generation but reappear in the second generation. So the factor that causes this characteristic must be kept separate, even in the tall first generation plants.

Mendel used pure-bred parents. He also:
• made the crosses by hand so he knew which parents produced which offspring
• repeated each test hundreds of times.

Now try this

Explain **two** ways in which Mendel's methods helped to make his results valid.   **(4 marks)**

Fossils

The fossil record provides evidence for the theory of **evolution** by **natural selection**.

Fossil formation

Fossils are found in rocks. They are the remains of organisms from millions of years ago. Fossils form in different ways, including:

1 when **traces** of organisms are preserved, such as burrows, rootlet traces and footprints (like the dinosaur footprints in the photo)

2 when parts of the organism, such as bones and teeth, are replaced by minerals as they decay

3 when soft tissues, such as muscles and leaves, do not decay because one or more of the conditions needed is missing.

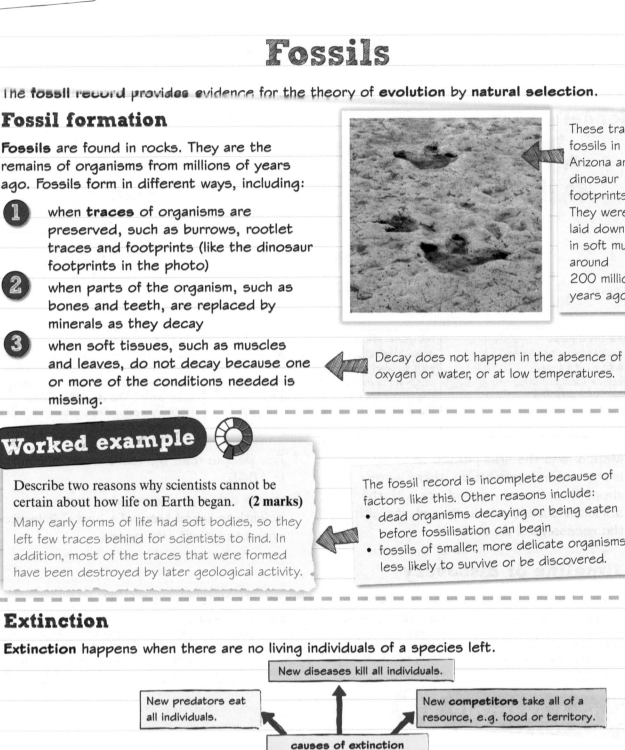

These trace fossils in Arizona are dinosaur footprints. They were laid down in soft mud around 200 million years ago.

Decay does not happen in the absence of oxygen or water, or at low temperatures.

Worked example

Describe two reasons why scientists cannot be certain about how life on Earth began. **(2 marks)**

Many early forms of life had soft bodies, so they left few traces behind for scientists to find. In addition, most of the traces that were formed have been destroyed by later geological activity.

The fossil record is incomplete because of factors like this. Other reasons include:
- dead organisms decaying or being eaten before fossilisation can begin
- fossils of smaller, more delicate organisms less likely to survive or be discovered.

Extinction

Extinction happens when there are no living individuals of a species left.

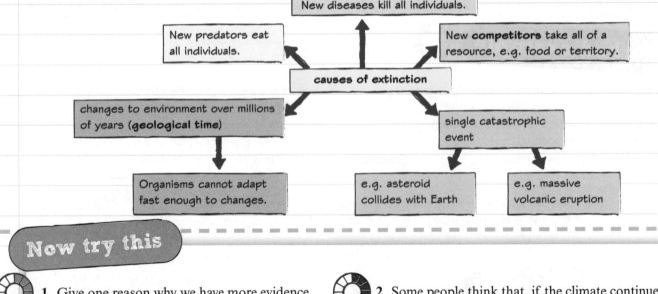

New diseases kill all individuals.

New predators eat all individuals.

New **competitors** take all of a resource, e.g. food or territory.

causes of extinction

changes to environment over millions of years (**geological time**)

single catastrophic event

Organisms cannot adapt fast enough to changes.

e.g. asteroid collides with Earth

e.g. massive volcanic eruption

Now try this

1 Give one reason why we have more evidence of how vertebrates (animals with bones) have evolved than of soft-bodied animals. **(1 mark)**

2 Some people think that, if the climate continues to change, Arctic animals such as the polar bear may soon become extinct. Explain why they may become extinct. **(2 marks)**

Resistant bacteria

Resistant strains

Bacterial strains resistant to an antibiotic can spread through a population of bacteria.

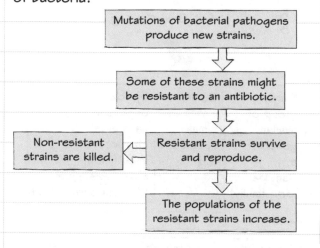

Mutations of bacterial pathogens produce new strains.

⬇

Some of these strains might be resistant to an antibiotic.

⬇

Resistant strains survive and reproduce. → Non-resistant strains are killed.

⬇

The populations of the resistant strains increase.

Multiple antibiotic resistance

The development of **antibiotic-resistant** strains of bacteria involves **natural selection**:
- The presence of an antibiotic provides an environmental condition that selects against bacteria that are killed or prevented from reproducing by the antibiotic.

Bacteria can evolve rapidly because they have a high rate of reproduction. A strain of bacteria that is resistant to one particular antibiotic may also develop resistance to another antibiotic:
- Mutations may produce individuals resistant to the second antibiotic.
- These individuals survive in both antibiotics.
- The population of this new strain increases.

MRSA

MRSA stands for meticillin-resistant *Staphylococcus aureus*:
- **Meticillin** (methicillin) is an antibiotic.
- *S. aureus* is a type of bacterium.

S. aureus is a common cause of skin and respiratory infections and food poisoning.

Meticillin is the antibiotic that used to be prescribed to treat *S. aureus* infections. It is not effective against MRSA. This causes problems in hospitals because:
- Patients may have wounds or weak immune systems due to illness.
- Doctors, nursing staff and visitors may transfer bacteria from place to place.

Reducing antibiotic resistance

We must reduce the development and spread of antibiotic-resistant strains because:
- people may not be immune to the bacteria
- there may be no effective treatment.

To reduce the rate that these strains form:

1 Doctors should not prescribe antibiotics to treat viral or non-serious bacterial infections.

2 Patients should finish antibiotic courses so all bacteria are killed, leaving none to form resistant strains.

3 The use of antibiotics in agriculture should be restricted.

Careful hygiene can reduce their spread.

> Antibiotics are used on farms to prevent infections and to increase growth rate.

Worked example

Give two reasons why the development of new antibiotics is not likely to keep up with the appearance of new resistant strains of bacteria. **(2 marks)**

It is expensive to develop new antibiotics, and the process takes a long time.

Now try this

Hospitals take care to reduce the spread of MRSA. Explain why:

(a) antiseptic gel dispensers are placed at entrances and near patients' beds **(2 marks)**

(b) in-patients have a swab taken from their skin or nose, for testing in a lab. **(2 marks)**

> About a third of people in the UK carry *S. aureus* without symptoms.

Classification

Classification involves grouping things in a systematic way according to their similarities.

Carl Linnaeus (1707–1778)

Linnaeus was a Swedish scientist. His **Linnaean system** classifies organisms into groups using their structure and characteristics:

- Kingdoms are the largest groups.
- **Species** are the smallest groups.

> A mnemonic like this can help you remember the groups in the right order:
> **K**ate's **P**oor **C**at **O**nly **F**eels **G**ood **S**ometimes.

Kingdom group of similar phyla, e.g. Animalia	Phylum group of similar classes, e.g. Chordata	Class group of similar orders, e.g. Mammalia	Order group of similar families, e.g. Carnivora	Family group of similar genera, e.g. Canidae (dog family)	Genus group of similar species, e.g. Canis	Species organisms that have most characteristics in common, e.g. *Canis lupus* (domesticated dog)

The binomial system

Every species has a unique **binomial** name. For example, the African lion is called *Panthera leo*.

genus name – shared with very similar species

species name – unique to African lions

Closely related organisms have the same genus name. For example:

- Leopards are *Panthera pardus*.
- Tigers are *Panthera tigris*.

Cheetahs are *Acinonyx jubatus*, showing that they are in a different genus and not closely related to lions, leopards and tigers.

Features of binomial names

Note that:

- the genus name starts with a capital letter but the species name does not
- typed binomial names are in *italics*
- written binomial names are <u>underlined</u>
- the genus name can be abbreviated to its first letter, e.g. *P. leo* for *Panthera leo*.

Binomial names are useful because:

- other people know exactly which species you mean
- you can see from the genus which species are very closely related (but modern DNA analysis can sometimes show that they are not).

Worked example

Explain why viruses are not in a kingdom. **(3 marks)**

Viruses do not have all the features of living organisms. Although they reproduce, they must use cells to do this. Viruses do not show other life processes such as respiration or growth.

> You can revise viral diseases on page 34.
> You may recall 'Mrs Gren' for life processes:
> - Movement
> - Respiration
> - Sensitivity
> - Growth
> - Reproduction
> - Excretion
> - Nutrition

Now try this

The common frog and viper (snake) are in the phylum Chordata. The viper and chameleon are in the class Reptilia. Explain whether the frog or the chameleon shares more characteristics with the viper. **(2 marks)**

Evolutionary trees

Developments in biology have had impacts on **classification systems**.

Three-domain system

Carl Woese proposed a **three-domain system** in 1977. This groups living things into three **domains** using evidence from chemical analysis:

 1 **Eukaryota**, which includes plants, animals, fungi and protists

 2 **Bacteria** – true bacteria and cyanobacteria (which can photosynthesise)

3 **Archaea** – primitive bacteria that usually live in extreme environments such as very hot or salty water.

Developing systems

The Linnaean system developed over time from two **kingdoms** (plants and animals) to six kingdoms or more. Simple organisms are more difficult to classify than complex ones.

New classification systems have been proposed because of improvements in:
- **microscopes**, increasing understanding of the internal structures of living organisms
- **chemical techniques**, increasing understanding of biochemical processes.

Evolutionary trees

An **evolutionary tree** is a way of showing the evolutionary relationships between species:
- Fossil data are used to work out relationships between extinct species.
- Modern classification data, including the results of DNA analysis, are used to work out relationships between living species.

Species B is the **common ancestor** of species C and D.

Species A is the **ancestor** of species B.

Species C and D are the **descendants** of species B.

going back in time   present

Worked example

The diagram shows an evolutionary tree.
(a) Describe what the point labelled **X** shows.
(2 marks)

X shows the last time at which the most recent common ancestor of humans and chimpanzees lived. This common ancestor is now extinct.

(b) Describe what the evolutionary tree shows about cows, whales and pigs.   **(2 marks)**

All three share a common ancestor. Cows and whales also share a more recent common ancestor. Both common ancestors are now extinct.

All six types of animal share a common ancestor in the distant past.

cow
whale
pig
hedgehog
human
chimpanzee

Now try this

Look at the evolutionary tree in the worked example.
(a) Describe what the tree shows about how similar humans, cows and pigs are.   **(2 marks)**
(b) Identify the pair of animals that shares the most recent common ancestor: cow and pig, or pig and hedgehog. Give a reason for your answer.   **(2 marks)**

95

Extended response – Inheritance, variation and evolution

There will be at least one 6-mark question on your exam paper. For these questions, you will need to think scientifically and structure your answer logically, showing how the points you make are related to each other.

You can revise the topic for this question, which is about inheritance, on page 83.

Worked example

Huntington's disease is an inherited disorder that damages nerve cells in the brain as the affected person gets older. It is caused by a dominant allele, H. Sickle cell disease is an inherited disorder that produces abnormally shaped red blood cells. It is caused by a recessive allele, s. Compare the inheritance of these two genetic disorders. You may use genetic diagrams to support your answer. **(6 marks)**

A person will have Huntington's disease if they inherit one or two copies of the dominant allele, H. On the other hand, a person will have sickle cell disease only if they inherit two copies of the recessive allele, s.

Two parents who are heterozygous for Huntington's disease have a 3 in 4 chance of producing a child with the disorder.

	H	h
H	HH affected	Hh affected
h	Hh affected	hh not affected

However, two parents who are heterozygous for sickle cell disease have only a 1 in 4 chance of producing a child who has the disorder.

	S	s
S	SS not affected	Ss carrier
s	Ss carrier	ss affected

Since sickle cell disease is caused by a recessive allele, people who are heterozygous are carriers. They do not have the disorder but can pass it on to their children. In contrast, Huntington's disease cannot have carriers because it is caused by a dominant allele.

> If you use genetic diagrams, you could use H and h for the alleles in Huntington's disease, and S and s for the alleles in sickle cell disease.

Command word: Compare

You need to describe the similarities and/or differences between things, not just write about one thing.

> This part of the answer explains an important difference in how people inherit these two disorders.

> A Punnett square diagram is a quick way to justify the prediction. Remember to indicate what each pair of alleles in the possible offspring means.

> Although the answer could discuss other genetic crosses, these two chosen crosses highlight a large difference in how the two disorders are inherited.

> Connectives such as 'in contrast', 'on the other hand' and 'however' are useful ways to show that a difference is being discussed.

Now try this

The inherited characteristics of a population can change over time because of natural selection or selective breeding. Compare these two processes. **(6 marks)**

> You can revise variation and evolution on page 84 and selective breeding on page 85.

Ecosystems

Levels of organisation

Ecosystems are organised at different levels.

organism		a single living individual
population		all the **organisms** of the same species in a habitat
community		the **populations** of all the different species in a habitat
ecosystem		the interaction of a **community** of living organisms with the non-living parts of their environment

A habitat is where an organism lives. The **environment** is all the conditions that surround an organism in its habitat. These are due to **abiotic** (non-living) factors and **biotic** (living) factors.

Abiotic factors

Abiotic (non-living) factors which can affect a community include:

- ☑ light intensity
- ☑ temperature
- ☑ moisture levels
- ☑ pH and mineral content of the soil
- ☑ intensity and direction of the wind.

In addition:
- oxygen levels are important to aquatic animals such as fish
- carbon dioxide levels are important for plants (as it is used in **photosynthesis**).

Biotic factors

Biotic (living) factors which can affect a community include:

- ☑ the availability of food
- ☑ whether new predators arrive
- ☑ whether new pathogens arrive
- ☑ **competition** between individuals of the same or different species.

Animals often compete with each other for:

- food   • mates   • territory.

One species may outcompete another species leaving too few individuals to breed.

Worked example

Describe three abiotic factors that plants in a community or habitat often compete for.

(3 marks)

Plants often compete with each other for light. They also often compete for water and mineral ions in the soil.

Plants also compete for space:
- below ground for their roots
- above ground for their shoots and leaves.

Light and water are needed for photosynthesis. Nitrate ions are needed to make proteins. You can revise photosynthesis on page 46.

Now try this

1 Describe the different levels of organisation in an ecosystem, from individual organisms to the whole ecosystem.   **(4 marks)**

2 The ground underneath the trees in a woodland is mostly bare. Explain why very few plants can grow there.   **(2 marks)**

Interdependence

If one species is removed from a community, it can affect the remaining species. This is called **interdependence**.

Stable communities

Organisms need materials from their surroundings and from other organisms so that they can survive and reproduce. Each species in a community depends on other species for resources, including food and shelter.

In a **stable community**:

👍 all the species and environmental factors are in balance, so

👍 the populations stay fairly constant in size.

Plant reproduction

In plants, male **gametes** (pollen cells) are transferred to female gametes (egg cells) by **pollination**. Some species of plants rely on the wind to carry their pollen from one plant to another. Other species rely on bees or other insects to do this.

Many species of plants rely on the wind to disperse their seeds far from the parent plants. Other species rely on animals to do this, for example when they eat fruits or carry seeds on their fur.

An example community

The diagram shows an interdependent community of plants and animals in a woodland habitat.

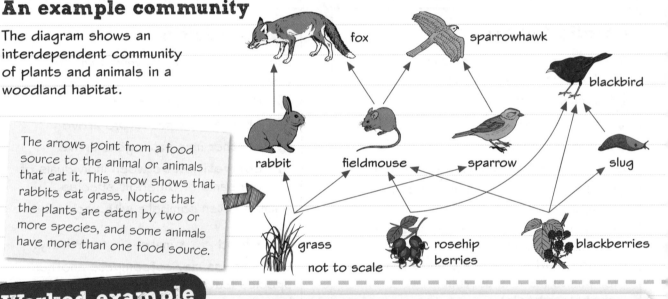

The arrows point from a food source to the animal or animals that eat it. This arrow shows that rabbits eat grass. Notice that the plants are eaten by two or more species, and some animals have more than one food source.

fox, sparrowhawk, blackbird, rabbit, fieldmouse, sparrow, slug, grass, rosehip berries, blackberries

not to scale

Now try this

Sparrowhawks are protected by law. However, a hunter illegally shoots these birds in the woodland. Explain **two** possible results of this new predator killing all the sparrowhawks. **(4 marks)**

Refer to the community in the diagram. What do sparrowhawks eat?

Adaptation

Organisms have features that let them survive in the conditions in which they normally live. These **adaptations** may be structural, behavioural or functional.

Cold environments

Animals in cold environments are **adapted** to reduce heat loss, and to move over snow.

small ears (reduced surface area) – less heat loss to air

white colour for **camouflage** against snow

large feet for better grip on ice and to stop bear sinking into snow

thick fur and fat below skin to **insulate** (reduce rate of heat loss)

Hot, dry environments

Animals in hot, dry environments are adapted to reduce heat gain and to maintain their water balance. For example, the large ears of African elephants transfer body heat to the air quickly. Camels have:

- 👍 thick fur at the top of the body that insulates against heat from the Sun, but thin fur elsewhere to allow heat loss
- 👍 the ability to tolerate high body temperatures and low body water content
- 👍 a hump of fat, which acts as a food store without insulating the rest of the body against heat loss.

Worked example

The diagram shows some features of a cactus plant. Explain how the cactus is adapted to survive in a dry environment. **(3 marks)**

The large root system allows the cactus to reach water deep underground. The thick fleshy body contains tissues that allow the cactus to store water. The green body allows the cactus to photosynthesise even though it has no leaves.

The leaves are adapted to form spikes and thorns to deter animals. You can revise plant defences on page 44.

green body without leaves

thick fleshy body with waterproof layer

large root system

Extreme environments

Some environments have extreme conditions. These include:

- high temperatures, e.g. volcanic springs
- high levels of salt, e.g. salt marshes
- high pressures, e.g. under the deep ocean.

Extremophiles

Extremophiles are organisms that live in very extreme environments. They have adaptations to let them survive in these environments.

Many extremophiles are microorganisms. For example, certain bacteria live in deep sea vents, where it is very hot, salty and under pressure.

Now try this

1 The Arctic hare lives where there is deep snow for many months, but the snow melts in summer. Explain why the hare has white fur in winter, but brown fur in the summer. **(2 marks)**

2 Many deep-ocean organisms can produce unique, flashing patterns of light. Suggest why they do this. **(2 marks)**

Think about the conditions at great depths.

Food chains

Feeding relationships within a community can be represented by **food chains**.

Producers and consumers

In a food chain, each arrow points from the organism being eaten to the organism that eats it. In the food chain in the diagram:
- rabbits eat grass
- foxes eat rabbits.

In general:
- producers are eaten by **primary consumers**
- primary consumers are eaten by **secondary consumers**
- secondary consumers are eaten by **tertiary consumers**.

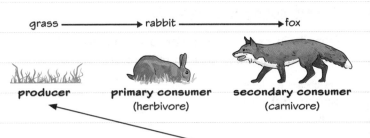

grass → rabbit → fox

producer | primary consumer (herbivore) | secondary consumer (carnivore)

All food chains begin with a **producer**, an organism that synthesises molecules. This is usually a green plant or alga. These make glucose by photosynthesis and produce **biomass** for life on Earth.

Predator–prey cycles

Predators are consumers that kill and eat other animals. **Prey** are the animals that are eaten.

In a **stable community**, the numbers of predators and prey rise and fall in cycles. In each cycle:

 The prey population rises if there is plenty of food and few predators. More prey survive to reproduce.

 As the prey population rises, there is more food for the predators. More of them survive to reproduce.

 As the predator population rises, they eat a greater proportion of prey, so the prey population falls.

4 As the prey population falls, there is less food for the predators, so their population falls too.

If the population size of the prey fell to zero, there would be no food for the predators. They would all starve and die unless they could find another food source.

The cycle repeats because the number of prey increases again:
- The remaining prey have more food.
- There are fewer predators to eat them.

The graph shows the population sizes of zebra and lions in a community. Explain the difference in the sizes of the two populations. **(3 marks)**

Lions are predators and zebra are their prey. A lion must eat several zebra each year to survive. This means that, although the numbers of lions and zebras change in cycles, there are fewer lions than zebra.

Start with the rise in the zebra population.

Explain the cycle of changes in population sizes shown in the Worked example graph. **(4 marks)**

Fieldwork techniques

Many methods are used to find the **abundance** and **distribution** of species in an ecosystem.

Abundance using quadrats

The most accurate way to find the size of a population is to count every individual. This is usually difficult, so samples are studied instead. A study place is sampled with a **quadrat** – a frame with a known area. To estimate a population size, you:

• Measure the area of the study place.
• Place quadrats randomly in the place.
• Count the number of study organisms inside each quadrat.

Quadrats are only useful for plants or very slow-moving animals. Random sampling is important because the organisms may not be evenly distributed in a study place.

Worked example

> Small quadrats are often 0.5 m × 0.5 m, giving an area of 0.25 m².

A rectangular school field measures 10 m × 20 m. Ten 0.25 m² quadrats are placed randomly. The number of daisy plants in each quadrat is counted. The table shows the results.

2	4	3	0	2	1	1	4	0	0

(a) Calculate the mean number of daisies in a quadrat. **(1 mark)**

mean number = 17/10 = 1.7

(b) Estimate the total population of daisy plants in the school field. **(2 marks)**

total population = 1.7 × 200/0.25 = 1360
area of school field = 10 × 20 = 200 m²

🖩 Maths skills Calculating a mean

To calculate a mean:
• Add the values together to get a total.
• Divide the total by the number of values.

🖩 Maths skills Mode and median

You can find these when you put the values in ascending order, e.g. for the daisy data:

0	0	0	1	1	2	2	3	4	4

The **mode** is the most frequently occurring value. It is 0 in these data (it occurs three times).

The **median** is the middle value in a range. If there is an even number of values (as here) you calculate the mean of the middle two values. This is (1 + 2)/2 = 1.5 here.

Distribution using transects

You can use a **transect** to study the effects of **abiotic** factors on the distribution of a species. You place the quadrats at regular intervals, not at random. You can find more details about this on page 102.

1 m × 1 m quadrats placed at regular intervals along the transect

transect line – e.g. tape measure placed along the ground

Now try this

Ten 1 m² quadrats were placed randomly on a 100 m by 200 m field. The number of dandelions recorded were: 20, 6, 33, 0, 26, 21, 18, 7, 2, and 9. Estimate the total number of dandelions on the field. **(3 marks)**

Field investigations

🧪 **Practical skills** You can use quadrats and transects to investigate the relationship between organisms and their environment in the field.

Core practical

Measuring a population size in a habitat

Aim

to estimate the population of a plant species using random sampling

Method

1 Use two tape measures to mark out and measure a large area of a field.
2 Record the number of dandelions inside at least 10 randomly placed 25 cm × 25 cm quadrats. Record your results in a suitable table.

Analysis

Use your results, and the marked area of the field, to estimate the population size.

Investigating the effect of a factor

Aim

to use a transect line to investigate the effect of light intensity

Method

1 Lay a tape measure in a straight line from the base of a tree to open ground.
2 Place a quadrat on the 0 m mark.
3 Count the number of dandelion plants inside the quadrat. Measure and record the light intensity using a light meter.
4 Repeat step 3 at regular intervals on the line.
5 Record your results in a suitable table.

Analysis

Plot a graph to show the results. This one shows some example results.

> Dandelion plants are easily recognised.

> You can revise this field technique on page 101. The area of a 25 cm quadrat is 0.0625 m² (0.25 m × 0.25 m).

> The estimated population size will be more accurate if you share results with other groups.

> You can revise how to do this on page 101.

> You could also investigate other abiotic factors such as soil pH and temperature.

> Take care not to lay your transect in dangerous places. The interval chosen is likely to differ between investigations. If it is:
> • small, the number of samples will be large and take time to study
> • large, you may miss important detail from your investigation.

> Remember to count plants, not flowers.

Now try this

(a) Describe how the number of dandelion plants changes with distance from the tree. **(1 mark)**
(b) Suggest an explanation for the distribution observed. **(3 marks)**

Cycling materials

The water cycle

The **water cycle** provides land plants and animals with fresh water.

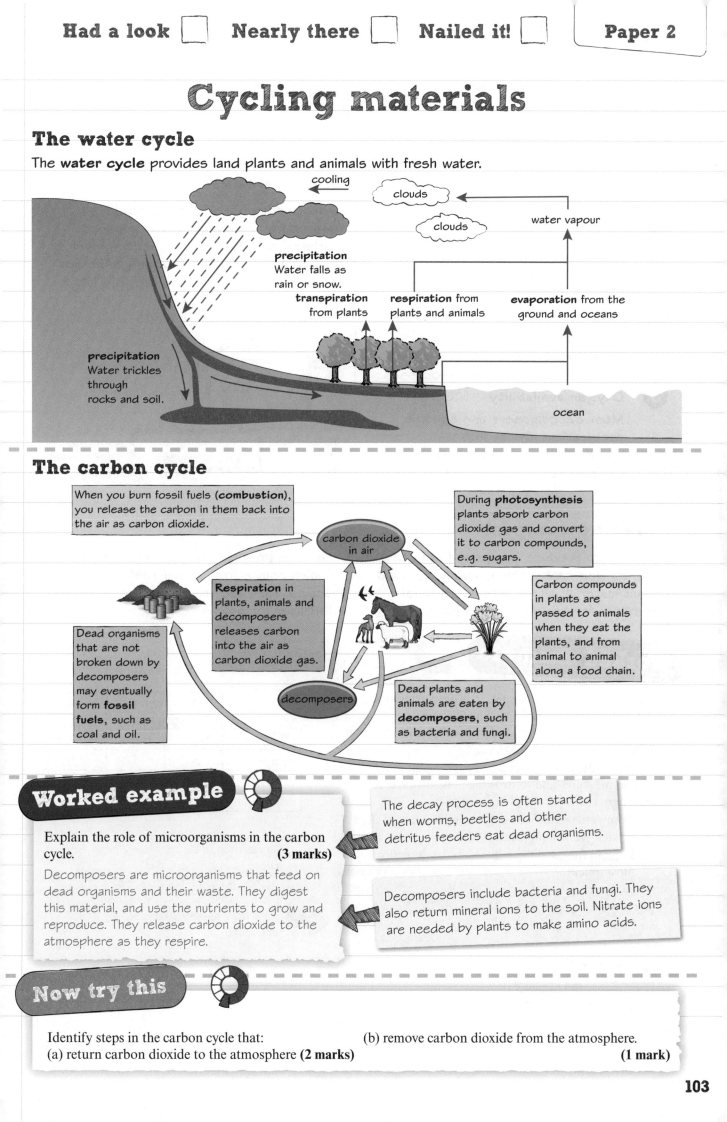

cooling

clouds

clouds

water vapour

precipitation
Water falls as
rain or snow.

transpiration
from plants

respiration from
plants and animals

evaporation from the
ground and oceans

precipitation
Water trickles
through
rocks and soil.

ocean

The carbon cycle

When you burn fossil fuels (**combustion**), you release the carbon in them back into the air as carbon dioxide.

During **photosynthesis** plants absorb carbon dioxide gas and convert it to carbon compounds, e.g. sugars.

carbon dioxide in air

Respiration in plants, animals and decomposers releases carbon into the air as carbon dioxide gas.

Carbon compounds in plants are passed to animals when they eat the plants, and from animal to animal along a food chain.

Dead organisms that are not broken down by decomposers may eventually form **fossil fuels**, such as coal and oil.

decomposers

Dead plants and animals are eaten by **decomposers**, such as bacteria and fungi.

Worked example

Explain the role of microorganisms in the carbon cycle. **(3 marks)**

Decomposers are microorganisms that feed on dead organisms and their waste. They digest this material, and use the nutrients to grow and reproduce. They release carbon dioxide to the atmosphere as they respire.

The decay process is often started when worms, beetles and other detritus feeders eat dead organisms.

Decomposers include bacteria and fungi. They also return mineral ions to the soil. Nitrate ions are needed by plants to make amino acids.

Now try this

Identify steps in the carbon cycle that:
(a) return carbon dioxide to the atmosphere **(2 marks)**

(b) remove carbon dioxide from the atmosphere.
(1 mark)

Decomposition

Decomposition is the breakdown or **decay** of biological materials.

Rate of decay

Factors affecting the rate of decay include:

 Temperature

Warm temperatures increase the rate of decay because they increase the rate of chemical reactions in microorganisms.

 Water content

Decay happens faster in moist conditions because microorganisms need water for their cell processes.

 Oxygen availability

Most decomposers use **aerobic respiration**. Decay is faster in the presence of oxygen because they can respire.

Biogas generators

Some decomposers use **anaerobic respiration**. Anaerobic decay produces **methane** gas. Methane is used as a fuel for cooking and heating. A **biogas generator** uses waste from animals to produce large volumes of methane.

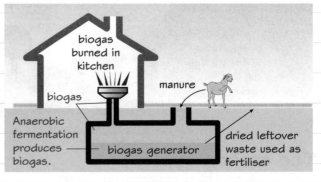

Making compost

Gardeners produce **compost** from kitchen and garden waste. This is mainly decayed plant material, and it acts as a natural fertiliser for growing garden plants.

Farmers produce compost and **manure** (which comes from animal waste) on larger scales. They use this material as a natural fertiliser for growing crops.

Worked example

A student investigates the decay of grass clippings. She sets up three insulated flasks. She fills one with damp grass, one with dry grass, and one with damp grass and disinfectant.

She places the flasks in the same room without lids. The student measures and records the mass of the green clippings. The table shows her results.

(a) Complete the table by calculating the rate of decay in each set of conditions. **(1 mark)**

(b) Explain the conditions in which the grass decayed most quickly. **(2 marks)**

The damp grass clippings decayed most quickly. This was because water and oxygen were available, and the flask maintained a warm temperature. These are the conditions in which decay microorganisms grow best.

Conditions	Mass of green clippings (g)		Rate of decay (g/day)
	Day 0	Day 6	
damp	50	5	7.5
dry	50	22	4.7
disinfectant	50	44	1.0

Dry clippings decay slowly because there is insufficient water for microorganisms to grow well. Disinfectant kills microorganisms, so decay happens very slowly in its presence.

Now try this

1 Explain why compost heaps produce compost most quickly in sunny places. **(2 marks)**

2 Suggest an explanation for why compost forms faster if it is turned over regularly. **(2 marks)**

Investigating decay

 Practical skills You can investigate the effect of temperature on the decay of milk by measuring pH change.

Core practical

Investigating the rate of decay

Aim

to investigate the effect of temperature on the rate of decay of fresh milk

Apparatus

- test tubes
- test-tube rack
- 10 cm³ graduated pipettes
- pipette filler
- thermometer
- stop clock
- water baths
- fresh full-fat milk
- lipase solution
- sodium carbonate solution
- cresol red indicator

Method

1 To a test tube, add 5 cm³ of milk, 7 cm³ of sodium carbonate solution and a few drops of cresol red indicator. Put the thermometer in the test tube.
2 Add about 5 cm³ of lipase solution to another test tube. Place both test tubes in a water bath and leave them until their temperature remains constant.
3 Add 1 cm³ of lipase solution to the milk mixture and start the stop clock. Stir until the indicator changes colour, and record the time taken.
4 Repeat steps 1 to 3 at different temperatures.

Results

Record your results in a suitable table.

Analysis

Plot a graph to show to show the results:
- time on the vertical axis
- temperature on the horizontal axis.

Natural decay of milk produces lactic acid, which causes the pH to fall. It may take several days for milk to decay or 'go off'. Lipase is used to model the decay of milk. The process is quicker and produces fatty acids, which also cause the pH to fall.

You may be given graduated plastic syringes instead of graduated pipettes.

Beakers of water from a kettle can be used if thermostatically controlled water baths are not available.

Sodium carbonate makes the mixture alkaline. Cresol red is purple in these conditions.

You could set up more than one test tube at step 1. This allows you to carry out repeats if you have time.

Cresol red is yellow in acidic conditions. You could use phenolphthalein indicator instead. This turns from pink in alkaline conditions to colourless in acidic conditions.

You can obtain temperatures below room temperature using an iced water bath. This would extend the range of temperatures investigated.

Now try this

A student carries out an investigation into the effect of temperature on the rate of decay of milk. Lipase solution is used to model the effects of decay bacteria.

(a) Explain why the time taken for the milk to become acidic decreases between 5 °C and 35 °C. **(2 marks)**
(b) Explain why the milk does not become acidic at 55 °C. **(3 marks)**

You can revise the effect of temperature on enzyme activity on page 19.

Environmental change

Changes that affect the distribution of a species include temperature and availability of water. These changes can be:

- seasonal changes
- geographical changes.

For example, it is colder in winter than it is in summer. Water is less available to plants and animals when it is frozen as ice or snow.

Geographical changes, such as the altitude, can also affect these and other factors.

In the photo, the wild flowers and trees grow at lower altitudes where it is warmer and water is more available. The surface is more likely to have soil than the mountainside.

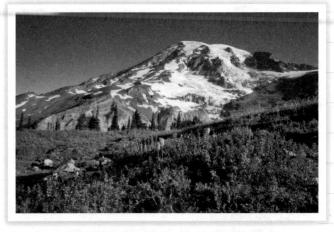

Summer in Mount Rainier National Park, America. Notice that there are far fewer trees higher up the mountain.

Seashores

Environmental changes happen regularly on seashores as the tide comes in and out. As you go higher up the shore:

✓ the temperature varies more

✓ the time out of water increases.

Periwinkles are a type of sea mollusc (snail):

✓ Small periwinkles are better adapted to the upper shore than rough periwinkles.

The diagram shows the distribution of these two species. The greater the height of the shaded areas in the two charts, the more periwinkles there are.

On this seashore, the rough periwinkles are distributed lower down the shore than the small periwinkles.

Worked example

Bushy lichens only grow when there is little air pollution from sulfur dioxide. The graph shows the results of a study in which the presence of these lichens on trees was studied. Suggest an explanation for the results. **(3 marks)**

The industrial area is a source of sulfur dioxide pollution. Bushy lichens could not grow there or nearby. The number of bushy lichens increases as the distance from the area increases. This is because the sulfur dioxide pollution decreases the further away from the area you go.

Human interactions also cause environmental changes, such as changing the composition of atmospheric gases. Sulfur dioxide may be produced when fuels such as diesel oil are burned.

Now try this

(a) Give **three** environmental changes that affect the distribution of species. **(3 marks)**

(b) Give the general causes of changes like the ones in your answer to part (a). **(3 marks)**

Waste management

Pollution kills animals and plants, which can lead to a reduction in **biodiversity**.

Pollution

Pollution is caused when a harmful or poisonous substance is released into the environment.

POLLUTION

Pesticides are used on crops to kill pests. **Herbicides** are used on crops to kill weeds.

of the air:
• smoke and gases, e.g. sulfur dioxide, which is part of acid rain

Sewage contains waste water as well as human waste.

of the land:
• toxic chemicals, e.g. pesticides and herbicides
• dumping waste in landfill sites

of the water:
• sewage
• fertiliser
• toxic chemicals washed in from land

Increasing pollution

The human **population** is growing rapidly.

rapid growth in human population + increased standard of living = increasing waste produced → if not handled properly, causes more pollution

uses more land for:
• building
• quarrying for building materials
• farming to produce food
• dumping waste
→ less space for other animals and plants

Worked example

Coal and oil release sulfur dioxide when they are burned. Natural gas releases much less of this toxic gas. In the UK, most sulfur dioxide emissions are from power stations. The graph shows total emissions between 2000 and 2014.

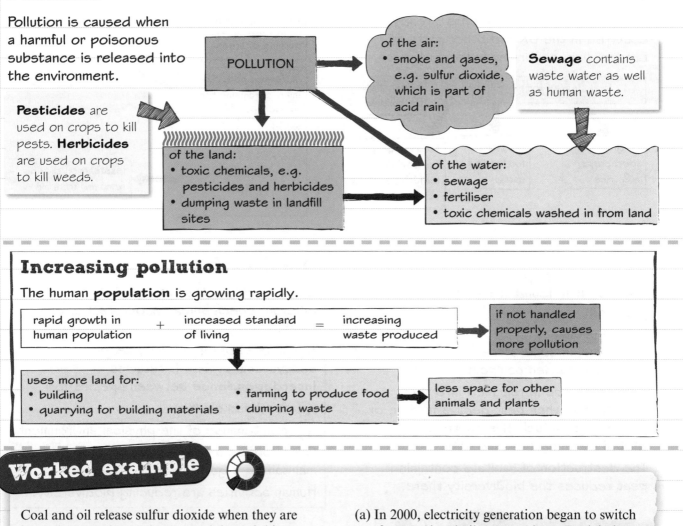

(a) In 2000, electricity generation began to switch from coal and oil to natural gas. Explain how the graph gives evidence of this. **(2 marks)**

Total emissions decreased after 2000. Natural gas releases less sulfur dioxide than coal and oil do, so switching to natural gas should reduce the amount released.

(b) More coal, which is cheaper than oil and gas, was used between 2010 and 2012. State how the graph gives evidence of this. **(1 mark)**

The total emissions were higher in 2010 and 2012 than they were in 2009.

Sulfur dioxide is a major cause of acid rain.

Now try this

Suggest an explanation for why pollution by pesticides and herbicides is increasing. **(3 marks)**

Deforestation

Deforestation is the permanent destruction of forests to make land available for other uses.

Reasons

Large-scale deforestation has already occurred in the UK and other parts of Europe. It is continuing in tropical areas.

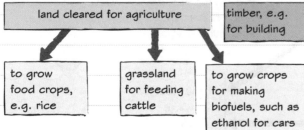

land cleared for agriculture

timber, e.g. for building

to grow food crops, e.g. rice

grassland for feeding cattle

to grow crops for making biofuels, such as ethanol for cars

Problems

Deforestation has environmental impacts.

burning of trees

increased activity of microorganisms that decay chopped wood

more carbon dioxide released into air

fewer trees so less carbon 'locked up' for many years in wood

less carbon dioxide removed from air

Peat

Peat consists of partly decomposed plant material. It is found in peat bogs and peatland.

Peat is useful:
- as a fuel in some places
- to make garden compost.

Carbon dioxide is released into the atmosphere when peat decays further, or is burned as a fuel. This contributes to **global warming** (see page 109).

The destruction of habitats containing peat reduces the biodiversity there.

Biodiversity

Biodiversity is the variety of all the different species in an ecosystem or on Earth.

A high level of biodiversity keeps ecosystems stable. It reduces **interdependence** between species for:
- ☑ food and shelter
- ☑ maintenance of the physical environment.

Our future as a species depends on us maintaining a good level of biodiversity.

Human activities are reducing biodiversity.

Worked example

Ground-living arthropods include beetles and millipedes. Palm oil comes from the fruit of oil palm trees. In parts of Borneo, areas of tropical rainforest have been cleared so oil palm trees can be planted instead. A study there found between 7 and 11 different arthropod groups in original rainforests. It found 4 or 5 arthropod groups in palm oil plantations.

Explain, using this information, the effect on the level of biodiversity of replacing rainforest with a single type of tree. **(2 marks)**

The study shows a reduced level of biodiversity in terms of the number of groups of arthropods. This is because fewer arthropod groups were found in the plantations than in the original rainforest.

Replacing different types of rainforest trees with oil palm trees reduces plant diversity.

A reduction in the number of arthropod groups will affect other organisms. Further studies might show a reduction in the number of other animals, such as orangutans.

Now try this

Explain why deforestation may lead to an increase in carbon dioxide in the air.
(3 marks)

You may also want to revise the carbon cycle on page 103.

Global warming

Greenhouse gases

Greenhouse gases such as methane and carbon dioxide absorb infrared radiation emitted by the Earth's surface, then release it in all directions. This helps to keep the Earth warm. Human activities are increasing the levels of greenhouse gases.

Greenhouse gas	Human activities
methane	farming (rice paddy fields and cattle)
carbon dioxide	combustion of coal, oil, natural gas and peat

Greenhouse effect

Increasing levels of greenhouse gases are causing an enhanced **greenhouse effect**. There is a scientific consensus that this is leading to **global warming** and **climate change**. This consensus is based on:

✓ thousands of peer-reviewed publications

✓ systematic reviews of these papers.

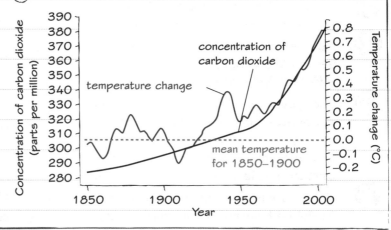

Effects of global warming

An increase in average temperatures of only a few degrees Celsius could have these effects:

• cause big changes in climate	• cause a rise in sea level	• affect species
e.g. drought, colder winters, hotter summers, flooding	due to ice melting and to warmer sea water (expands as it gets warmer)	e.g. change in migration of birds, change in distribution of species (the area they live in)

Worked example

Bluebell plants produce flowers in spring. The date of the first bluebell flower in a particular woodland was recorded between 1996 and 2012. The table shows the results.

Year of study	1996	2000	2004	2008	2012
Date of first bluebell flower	23 April	13 April	14 April	10 April	4 April

(a) Explain one way in which the method used improves the reliability of the results. **(2 marks)**

The data were collected at a single site, rather than from two or more sites. This helps to control factors that differ between sites.

Data from a single species may not lead to a valid conclusion.

(b) Explain whether the data provide evidence for climate change. **(2 marks)**

Overall, the flowering date becomes earlier during the study. This suggests that the climate is changing, for example becoming warmer earlier in the year. However, we do not know what happened in the years between each visit.

Now try this

Suggest how climate change may make it possible for tropical diseases to spread to new areas. **(3 marks)**

Maintaining biodiversity

Many human activities reduce **biodiversity**. Steps have only been taken recently to stop this.

Reasons for positive action

There are programmes to reduce the negative impact of human activities on the environment. These have been put in place by scientists and other concerned people. Reasons include:

1 Moral and ethical reasons: humans should respect other species.

2 Aesthetic reasons: people enjoy seeing the variety of species in different **habitats**.

3 Ecosystem structure: organisms in an ecosystem are **interdependent** (revise this on page 98). For example, decay microorganisms are important in the **carbon cycle** (revise this on page 103).

4 Value: some species are particularly useful to humans. For example, plants may be a source of new medicines (revise this on page 41), and others provide opportunities to make money from tourism.

Some programmes

Programmes to reduce or reverse the decline in biodiversity include:

- ✓ protecting and regenerating rare habitats, such as wetlands
- ✓ replanting hedgerows in farms
- ✓ recycling resources rather than dumping waste in landfill, which reduces activities such as quarrying
- ✓ reducing deforestation.

Replanting forests

Advantages of **reforestation** include:

- ✓ restores habitat for endangered species
- ✓ reduces the effects of soil erosion because tree roots bind soil together
- ✓ helps to reduce the overall release of carbon dioxide as the trees photosynthesise.

Worked example

The graph shows the changes in the Asian tiger population.

Population of tigers (y-axis: 0 to 40 000). x-axis: 1970 to 2015.

Extinction happens when no living individuals of a species remain (you can revise causes of this on page 92).

An **endangered species** faces a high risk of extinction in the near future. There are **breeding programmes** for some of these species. Animals or plants are bred over several generations. Reproduction is planned to increase numbers without causing **inbreeding** (you can revise this on page 85).

(a) During this time, the human population in tiger regions increased significantly.
Explain how this might be linked to the size of the tiger population. **(2 marks)**

The tiger population decreased as the human population increased. Human activities such as farming and building may be destroying the tigers' habitat. Hunters may be killing tigers.

(b) Scientists may closely monitor the population size of large carnivores like tigers when they assess the biodiversity of an area. Suggest a reason for this. **(2 marks)**

Large carnivores eat other animals, so they may rely on complex food chains. If there are many large carnivores in an area, it suggests that there may be many other species there too.

Now try this

Explain how planting hedges around fields can help maintain biodiversity. **(3 marks)**

Trophic levels

Trophic levels are feeding levels along a **food chain**, starting at level 1, the producers.

Pyramids of biomass

Biomass is the mass of material in an organism.

It is often given as the dry mass – the mass of the organism without its water.

A **pyramid of biomass** shows the relative amount of biomass at each trophic level.

You can revise food chains on page 100.

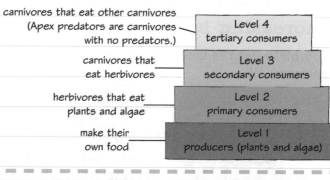

carnivores that eat other carnivores (Apex predators are carnivores with no predators.)

carnivores that eat herbivores

herbivores that eat plants and algae

make their own food

| Level 4 tertiary consumers |
| Level 3 secondary consumers |
| Level 2 primary consumers |
| Level I producers (plants and algae) |

The table shows the biomass of organisms in the food chain: lettuce → caterpillar → thrush

Organism	Biomass (g/m²)
lettuce	120
caterpillar	60
thrush	12

(a) Use the data to draw an accurate pyramid of biomass. **(2 marks)**

(b) Give one reason why large amounts of glucose are lost at each trophic level. **(1 mark)**

The glucose is used for respiration.

(c) Explain the shape of the pyramid of biomass. **(3 marks)**

Biomass is lost between one trophic level and the next. This is because the caterpillars and thrushes cannot digest and absorb all the material they eat. Some of the biomass eaten by the caterpillars is egested as faeces, so is not eaten by the thrushes.

Draw the bars to scale, so the width of each one represents the biomass given to you.

thrush / caterpillar / lettuce

Remember that all living organisms respire, so glucose is used for respiration in plants **and** animals.

About 1% of the energy from light received by plants and algae is transferred for photosynthesis.

In general, only about 10% of the biomass in one trophic level is passed to the level above.

Other biomass losses include:
• carbon dioxide and water in respiration
• water and urea in urine.

Maths skills — Efficiency

60 g of caterpillar becomes 12 g of thrush. The **efficiency** is:

$$\frac{12}{60} = \frac{1}{5}$$

Multiply by 100 to obtain a percentage:

$$100 \times \frac{1}{5} = 20\%$$

Numbers

The numbers at each trophic level depend on:
- ☑ the biomass of each individual
- ☑ the total biomass of the trophic level.

For a given total biomass, small organisms will be greater in number than large organisms.

A rabbit eats 500 g of grass. Give **four** reasons why the rabbit will not increase in biomass by 500 g. **(4 marks)**

Food security

Food security is having enough food to feed a population. **Sustainable** methods of food production must be found to feed everyone on Earth.

Increasing birth rate

The **birth rate** is increasing in parts of the world. This affects food security and the environment.

| Increasing **human population** means more food is needed. | → | The increasing demand for meat and fish means more land is used for **animal farming** and a greater impact on wild fish populations. | → | Movement of people and goods introduces new pests and pathogens to areas, damaging local crops and animals. | → | Increased waste is produced, causing more pollution. |

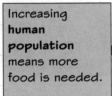 **Worked example**

Bluetongue is a viral disease of sheep and cattle. It is spread by insects. The disease causes a serious illness and many infected animals will die from it. Bluetongue is usually found in warm parts of southern Europe.

(a) Bluetongue has recently spread to the UK, as shown on the map. Suggest an explanation for why scientists think this may be due to climate change. **(2 marks)**

The insect vector that spreads bluetongue may need a warm climate to survive. If climate change is making the UK warmer, it is likely that these insects can now survive in the UK.

(b) Give one reason why farmers in the UK are worried about bluetongue. **(1 mark)**

The disease will make farm animals ill or even kill them. This would reduce the food supply.

When new diseases or pests spread to an area, the existing plants or animals may not have resistance against them. This can seriously affect farming, reducing the yield and quality of the food produced, and the farmers' profits.

Other factors affecting food security

Conflicts can also reduce the availability of fresh water for drinking and for watering crops. →

Wars and other conflicts make it difficult for farmers to grow, process and transport food.

increased cost of fuel, fertilisers, machinery, seed, animal feed

reduced food supply

changing diets in developed countries:
• some food supplied by less well developed countries

environmental changes, such as reduced rainfall:
• may cause widespread famine

 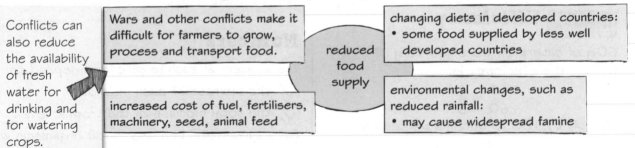

Now try this

About 80% of the population of India are Hindu, and vegetarianism is widespread. Cows are sacred in Hindu culture. However, India is the world's largest exporter of beef. This is because buffalo, which supply India's huge dairy industry, are classed as beef.

Explain whether buffalo meat exports are likely to affect food security in India. **(3 marks)**

Farming techniques

Farmers may use **intensive farming** techniques to increase the yield of food. This type of farming is sometimes called 'factory farming'.

Intensive farming

The efficiency of meat production can be improved by:

- keeping animals in pens or cages to limit their movement
- keeping the surroundings warm.

This reduces energy transfer from the animals to their environment:

Animals may also be given diets high in protein to increase their growth.

> Food animals lose energy:
> - by moving around
> - as heat to the environment.
>
> ↓
>
> Less energy is lost if:
> - movement is restricted, e.g. animals are kept in cages
> - the environment is kept warm.
>
> ↓
>
> ✓ • The animals grow faster and bigger.
> ✗ • Animals close together are more likely to get ill.
> ✗ • Heating uses electricity.
> ✗ • It is unethical to keep animals in stressful or unhealthy conditions.

Worked example

Free-range chickens can move around freely, flap their wings, and behave naturally. Intensively farmed chickens have less room to move around and may be kept in large, heated barns. Free-range chicken is usually more expensive to buy than intensively farmed chicken.

Evaluate the use of these two methods to produce chicken for human consumption. **(5 marks)**

> You do not need to give a conclusion unless you are asked to do so.

The free-range chickens may grow more slowly than the intensively farmed chickens. This is because they will lose more energy keeping warm and moving around, so they will use more glucose for respiration. On the other hand, they can behave more naturally than the intensively farmed chickens. People may be happy to pay more money for free-range chicken because they will think it has been produced more ethically. However, some people may not be able to afford free-range chicken.

Plant diet vs animal diet

More food is available for humans from the same mass of plant material if we eat the plants, rather than eating meat from animals that eat the plants.

Biomass is lost from one trophic level to the next (you can revise this topic on page 111).

Now try this

Refer to this food chain. In terms of efficiency of food production, explain at which trophic level humans should gather food from the sea.

microscopic plankton → sardine (small fish) → tuna (large fish)   **(4 marks)**

Sustainable fisheries

People catch fish for food. Fish stocks must be maintained at levels that allow breeding to occur, or some species of fish may disappear in some areas.

Catching fish

Fishermen catch many sizes of fish. If there are not enough large fish in the catch, they may also take smaller fish. The rest of the catch is usually thrown back into the water.

> It takes several years for fish to mature and become large enough to breed.

Only fish bigger than this can breed.

Bigger fish are more valuable than smaller fish.

A shoal contains fish of different sizes and ages.

Reasons for declines in fish stocks

| If fishing takes all the large fish ... | → | there are no fish left that are big enough to breed ... | → | so no young fish replace the fish taken in fishing ... | → | the fishery collapses (there are not enough fish to make fishing worthwhile). |

In a **sustainable fishery**, we could catch enough fish for our needs without affecting the needs of people in the future. The populations of the different species would be stable. However, if we catch too many fish to be sustainable (we overfish), we will have:

- ✓ plenty of fish in the short term
- ✓ few fish or no fish in the longer term.

Worked example

Fish stocks must be conserved to make a fishery sustainable. Fishing quotas and limits on mesh size can achieve this or allow stocks to recover.

(a) Explain what is meant by a fishing quota. **(2 marks)**

A fishing quota is a maximum number or mass of fish that fishermen are allowed to catch. Quotas usually apply to a particular species of fish in an effort to let stocks increase.

> Another way to allow stocks to recover is to close an area to fishing for several years.

(b) Explain how limits on the mesh size of fishing nets can help to conserve fish stocks. **(3 marks)**

The mesh size is limited to a minimum size. This ensures that small fish are not caught. If the mesh limit is larger than the smallest size for a breeding fish, there will always be some fish left in the area that can reproduce.

> Quotas and other conservation measures are often unpopular with people in the fishing industry because employment is reduced.

Now try this

Before a minimum mesh size was introduced, the European Union set a minimum size of fish that could be sold, or caught in its fishing areas.

(a) Suggest a reason why this led to fishermen dumping dead fish overboard before they returned to port. **(1 mark)**

(b) Explain how setting a certain mesh size helped to prevent this happening. **(2 marks)**

Biotechnology and food

Mycoprotein

Fusarium is a type of fungus. It is useful for producing **mycoprotein**. This is:

• rich in protein

• suitable for vegetarians.

Fusarium is grown in large amounts in huge containers called fermenters. These are around 50 m high (see diagram). In the fermenter:

1 Air is added to provide aerobic conditions so the fungal cells can respire.

2 Glucose syrup is added as a source of food for the fungal cells.

3 The culture mixture or 'broth' continually circulates. *Fusarium* is dense and sinks to the bottom.

4 *Fusarium* is harvested and purified. It can be shaped into pieces to imitate the appearance and texture of meat.

Ammonia is added as a source of nitrogen for the *Fusarium* cells to make proteins.

Rising bubbles cause broth to circulate.

Denser broth falls to base of fermenter.

gases removed

ammonia and air

glucose syrup and minerals

heat treatment

cooling system

drying and chilling

mycoprotein harvested

Genetic modification

Genetic engineering is used to produce **genetically modified** (GM) organisms:

☑ GM bacteria produce human insulin to treat people with diabetes.

☑ GM crops with high yields can provide more food.

☑ GM crops such as 'golden rice' can provide improved nutritional value.

You can revise genetic engineering on page 86 and details of how it works on page 87.

☑ Making human insulin using GM bacteria is quicker and cheaper than producing it any other way.

☒ A few diabetic people react badly to this insulin and need a different form.

☑ Golden rice has genes that increase its production of β-carotene, needed to make vitamin A. Eating golden rice could prevent vitamin A deficiency.

☒ Golden rice seed costs more than normal rice seed.

Worked example

The molecules are absorbed by diffusion.

Most mushrooms are grown commercially. Explain why they can be grown on a variety of substances, including wood chips and manure. **(3 marks)**

Mushrooms are decomposers. Like other decomposers, they secrete enzymes into the environment that break down dead plant and animal material. They then absorb the small, soluble molecules formed.

Now try this

Describe **two** ways in which biotechnology can increase the amount of food for people. **(4 marks)**

Extended response – Ecology

There will be at least one 6 mark question on your exam paper. For these questions, you will need to think scientifically and structure your answer logically, showing how the points you make are related to each other.

You can revise the topics for this question, which is about deforestation and global warming, on pages 108 and 109.

Worked example

Large-scale deforestation has occurred in tropical areas. It has environmental implications, including changing the amounts of some gases in the atmosphere. These changes cause global warming, possibly leading to climate change.

Explain why deforestation is taking place, and describe how it causes changes to the atmosphere. **(6 marks)**

The human population is growing and standards of living are increasing. More land is needed to sustain these changes. Deforestation involves clearing forests to provide more land for human activities.

The land released by deforestation may be used to grow crops for biofuels, such as ethanol for cars. It is also used to provide grazing for cattle, or fields to grow rice. This allows more food to be produced. However, it also leads to an increase in methane in the air, because cattle and rice fields release this gas.

Deforestation leads to an increase in carbon dioxide in the atmosphere. The trees may be burned after they are cut down, releasing carbon dioxide. Decomposers break down wood chips and branches, and they release carbon dioxide by respiration. Photosynthesis uses carbon dioxide from the atmosphere. If there are fewer trees to photosynthesise, less carbon dioxide will be absorbed.

> **Command word: Explain**
>
> If you are asked to **explain** something, you need to make it clear, or state the reasons why it happens. You should not just give a list of reasons.
>
> **Command word: Describe**
>
> If you are asked to **describe** something, you need to recall some facts, events or processes in an accurate way.

This part of the answer gives a general explanation of why deforestation happens.

Three detailed reasons for deforestation are given here. Two of these reasons are linked to an increase in methane in the atmosphere.

Trees and other plants 'lock up' carbon dioxide as carbon compounds in their tissues. Two main effects of deforestation are described. These are how carbon dioxide is released, and how less carbon dioxide is absorbed. Both lead to an increase in levels of this gas in the atmosphere.

You need to organise your answer into a logical order, with ideas linked in a sensible way. Take care to include only the information required by the question. In this case, although increases in carbon dioxide and methane are a cause of global warming and climate change, there is no need to describe or explain these processes.

Now try this

The factory farming of animals is one way to increase the efficiency of food production. Evaluate the advantages and disadvantages of the factory farming of pigs to provide meat for human consumption. **(6 marks)**

You can revise farming techniques on page 113. Remember that antibiotics may be used in agriculture (revise the problems this causes on page 93).

Answers

> **Extended response questions**
>
> Answers to 6-mark questions are indicated with a star (*).
> In your exam, your answers to 6-mark questions will be marked on how well you present and organise your response, not just on the scientific content. Your responses should contain most or all of the points given in the answers below, but you should also make sure that you show how the points link to each other, and structure your response in a clear and logical way.

1. Microscopes and magnification

total magnification = 5 × 20 = ×100 **(1)**

actual length = 15/100 = 0.15 mm **(1)**

actual length = 0.15 × 1000 = 150 μm **(1)**

2. Animal and plant cells

1 cell membrane **(1)**; nucleus **(1)**; cytoplasm **(1)**; mitochondria **(1)**; ribosomes **(1)**

2 Muscle cells respire more/need more energy. **(1)** Mitochondria are where respiration happens/energy is released **(1)**.

3 Plant roots are in the dark **(1)** so they do not carry out photosynthesis **(1)**.

3. Eukaryotes and prokaryotes

(a) (i) 7.5×10^{-5} m **(1)**

   (ii) 7.5×10^{-7} m **(1)**

(b) two orders of magnitude **(1)**

4. Specialised animal cells

Ciliated cells are rectangular/column shaped **(1)** so they can fit together to line the trachea **(1)**.

Sperm cells are streamlined/long and thin **(1)** to help them swim more easily **(1)**.

5. Specialised plant cells

a long thin finger-like part **(1)** that grows between soil particles **(1)**

6. Using a light microscope

neat pencil drawing without unnecessary shading **(1)**, cell wall labelled **(1)**, nucleus labelled **(1)**, cytoplasm labelled **(1)**, chloroplast labelled **(1)**, e.g.

7. Aseptic techniques

1 gamma radiation **(1)**; heating in an autoclave **(1)**; heating in a flame **(1)**

2 This is to kill bacteria on the bench **(1)** that could contaminate the experiment/cultures/to prevent contamination **(1)**.

3 4 hours = 4 × 60 = 240 minutes, so number of divisions = 240/40 = 6 **(1)** or

40 minutes = 40/60 = 0.667 hours, so number of divisions = 4/0.667 = 6 **(1)**

number of bacteria = 2500 × 2^6 = 2500 × 64 = 160 000 **(1)**

8. Investigating microbial cultures

(a) antiseptic B: mean diameter = 25 mm **(1)**; area = 4.91×10^{-4} m^2 **(1)**

antiseptic C: mean diameter = 20 mm **(1)**; area = 3.14×10^{-4} m^2 **(1)**

(b) Antiseptic B kills the most bacteria **(1)** because it has the largest clear zone **(1)**.

9. Mitosis

DNA is replicated (at the end of interphase) **(1)**, forming two copies of each chromosome **(1)**; during mitosis (and cell division) a cell divides in two/one set of chromosomes goes into each new cell **(1)**.

10. Stem cells

Advantage, e.g. stem cells are genetically identical to the patient's cells/contain the same genes as the patient **(1)**, so the cells will not be rejected by the patient's body **(1)**.

Disadvantage, e.g. egg cell used/embryo created **(1)** and people may have ethical/religious objections to this **(1)**.

11. Diffusion

1 More particles (of a gas or a dissolved substance) **(1)** move from an area of high concentration to an area of low concentration/down a concentration gradient **(1)** so causing a net/overall movement of particles **(1)**.

2 The rate of diffusion increases **(1)** because the particles move more quickly **(1)**.

12. Exchange surfaces

1 having an efficient blood supply **(1)**; keeping the exchange surface ventilated **(1)**

2 Single-celled organisms have a relatively large surface area to volume ratio **(1)** so molecules move in and out sufficiently well to meet their needs **(1)**; but multicellular organisms have smaller surface area to volume ratios **(1)** so exchange surfaces increase the surface area available for exchange of materials **(1)**.

13. Osmosis

1 the diffusion of water **(1)** from a dilute solution to a concentrated solution **(1)** through a partially permeable membrane **(1)**

2 Sugar dissolves in water from the fruit **(1)** producing a concentrated sugar solution **(1)**, so water continues to leave the fruit by osmosis **(1)**.

14. Investigating osmosis

(a) 1 mark for each correct percentage change in mass:

Concentration of sucrose (mol/dm³)	Change in mass (g)	% change in mass
0.25	+0.08	+3.7
0.50	−0.25	−12.2
0.75	−0.43	−21.9

(b) axes labelled with quantity and unit, and scales chosen so plotted points cover at least 50% **(1)**; all five points plotted ± ½ square **(1)**; curve of best fit is drawn **(1)**; e.g.

(c) Concentration is estimated from graph at 0% change, e.g. 0.3 mol/dm³. **(1)**

15. Active transport

Mitochondria release energy by respiration in cells **(1)**; cells involved in active transport need a lot of energy (so they have many mitochondria) **(1)**.

16. Extended response – Cell biology

*Answer could include the following points to 6 marks:

Similarities (both have):

- nucleus
- cytoplasm
- cell membrane
- mitochondria
- ribosomes.

Differences:

- Plant cells have a cell wall.
- Cell wall is made from cellulose.
- Plant cells often have chloroplasts.
- Plant cells often have a permanent vacuole.
- Vacuole is filled with cell sap.
- Plant cells often have regular shapes.

Functions:

- nucleus – contains genes/controls cell activities
- cytoplasm – site of many reactions that take place in the cell
- cell membrane – controls what substances enter or leave the cell
- mitochondria – where respiration happens/release energy for the cell
- ribosomes – site of protein synthesis/where proteins are made
- cell wall – strengthens/supports the cell
- chloroplasts – site of photosynthesis/where photosynthesis happens
- vacuole – helps to keep the cell rigid.

17. The digestive system

suitable summary table, e.g.

	Amylase	Proteases	Lipases
Sites of production	salivary glands pancreas small intestine	stomach pancreas small intestine	pancreas small intestine
Sites of action	mouth small intestine	stomach small intestine	small intestine
Substances broken down	starch	proteins	lipids/fats/oils
Products formed	simple sugars	amino acids	fatty acids and glycerol

1 mark for each correct site of production *and* action for each enzyme, to 3 marks.

1 mark for each correct substance broken down *and* products for each enzyme, to 3 marks.

18. Food testing

It is safer to heat a test tube of liquid using a hot water bath as liquid may boil or escape if a Bunsen burner is used **(1)**; the Sudan III stain is dissolved in ethanol **(1)** which is highly flammable/its vapour could ignite if a Bunsen burner flame is nearby **(1)**.

19. Enzymes

1 pH value given <7 **(1)**; the conditions in the stomach are acidic/stomach produces hydrochloric acid **(1)**

2 The shape of the active site must match the shape of the substrate/reactant **(1)**; the shape of the active site changes as the temperature and pH change **(1)**. The optimum temperature/optimum pH is when the activity of the enzyme is greatest **(1)**; the activity of the enzyme is lower below and above this temperature/pH **(1)**.

20. Investigating enzymes

1 correct values **(1)** given to 2 significant figures **(1)**

pH of buffer	Time taken to digest starch (s)	Rate of reaction (/s)
4	169	0.0059
5	125	1/125 = 0.0080
6	100	1/100 = 0.010
7	118	1/118 = 0.0085
8	260	1/260 = 0.0038

2 axes labelled with quantity and unit, and scales chosen so plotted points cover at least 50% **(1)**; all five points plotted ±½ square **(1)**; curve of best fit drawn **(1)**; e.g.

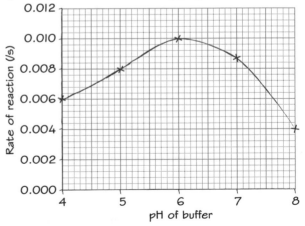

3 The optimum pH is 6 **(1)**; this is when the rate of reaction is greatest/the reaction time is the lowest **(1)**.

4 Temperature also affects enzyme activity/rate of reaction **(1)** so it must be kept the same to make sure the results are valid **(1)**.

21. The blood

Red blood cells are smaller than white blood cells **(1)**; red blood cells do not have a nucleus but white blood cells do **(1)**; red blood cells have a biconcave shape but white blood cells do not **(1)**; red blood cells contain haemoglobin but white blood cells do not **(1)**.

22. Blood vessels

1 artery, capillary, vein **(1)**

2 Cells need oxygen and glucose for respiration, and to get rid of waste products/carbon dioxide **(1)**; capillaries are the blood vessels that exchange substances with the body cells **(1)**; a short distance between a body cell and a capillary allows efficient/fast diffusion **(1)**.

3 to allow blood/blood cells to move easily/smoothly through the artery **(1)**

23. The heart

The right ventricle pumps blood to the lungs, but the left ventricle pumps blood to the rest of the body **(1)**, so the left ventricle needs to pump with greater force than the right ventricle **(1)**; the lungs/alveoli/capillaries in the lungs would be damaged by very high pressure blood **(1)**.

24. The lungs

Exhaled air has less oxygen **(1)**, more carbon dioxide **(1)** and more water vapour **(1)**.

25. Cardiovascular disease

any two from: risk of complications **(1)**; surgery is expensive **(1)**; patient may take time to recover after operation **(1)**; risk of infection **(1)**

26. Health and disease

It is a non-communicable disease because it is not caused by a pathogen **(1)** and blood clots are not spread from one person to another **(1)**.

27. Lifestyle and disease

unfit men of normal weight/fit men of normal weight **(1)**

28. Alcohol and smoking

1 brain damage **(1)**; liver damage/cirrhosis **(1)**

2 Substances in smoke narrow the blood vessels **(1)**, increasing blood pressure **(1)**.

29. The leaf

Chloroplasts are needed for photosynthesis **(1)**; chloroplasts are found in guard cells but not in other epidermis cells **(1)**.

30. Transpiration

1 Water enters root hair cell by osmosis **(1)**, moves through plant via xylem **(1)**, evaporates from leaf as water vapour **(1)** via stomata **(1)**.

2 Stomata open when guard cells become turgid **(1)** but close when guard cells become flaccid **(1)**, so more water vapour is lost in the light/during the day **(1)** when the stomata are open **(1)**.

31. Investigating transpiration

1 light intensity **(1)**; temperature **(1)**; air movement **(1)**; humidity **(1)**

2 At higher temperatures, the rate of photosynthesis is greater **(1)** so more stomata open **(1)**; water molecules move more quickly **(1)** and when the air moves faster (past the leaves) the concentration gradient of water vapour increases **(1)**, so the rate of diffusion increases **(1)**.

32. Translocation

Photosynthesis does not happen in the dark **(1)**; starch/storage substances broken down **(1)** to form sugars/glucose **(1)**; phloem can transport dissolved sugars/glucose to the rest of the plant **(1)**.

33. Extended response – Organisation

*Answer could include the following points to 6 marks:

Surface area:
- millions/many alveoli in the lungs
- increases surface area
- Larger surface area gives greater rate of diffusion.

Diffusion distance:
- Walls are thin.
- one cell thick
- provides a short diffusion distance
- Shorter diffusion distance gives greater rate of diffusion.

Difference in concentration:
- Inhaled air has a high concentration of oxygen/higher concentration than the blood.
- Large difference in concentration/high concentration gradient gives greater rate of diffusion.
- Blood carries oxygen away from alveoli.
- Maintains high concentration gradient.
- Blood has high concentration of carbon dioxide/higher concentration than inhaled air.
- Large difference in concentration/high concentration gradient gives greater rate of diffusion.
- Breathing out carries carbon dioxide away from alveoli.
- maintains high concentration gradient.

34. Viral diseases

1 two from the following for 1 mark each: direct contact; by water; by air

2 Viruses exist/live/reproduce in cells **(1)**; burst out/escape from the cell **(1)**.

35. Bacterial diseases

1 They damage tissues/make us feel ill. **(1)**

2 Washing hands kills/removes bacteria **(1)**, stopping bacteria spreading/getting on the food **(1)**.

36. Fungal and protist diseases

1 Use fungicide to kill the fungi **(1)**; wash contaminated clothing/avoid skin contact with an infected person/avoid contact with contaminated clothing **(1)**.

2 (a) When an infected mosquito injects the malaria parasite into a person **(1)** the parasite will be killed (before it can infect the liver cells) **(1)**.

   (b) There will be no malaria parasites in the blood **(1)** so when a mosquito bites/feeds it will not become infected **(1)**.

   (c) There will be fewer places for mosquitos to breed **(1)**, which reduces the chance of malaria spreading/there will be fewer disease vectors **(1)**.

37. Human defence systems

1 any two from the following for 1 mark each: skin/scabs; sticky mucus; cilia

2 Saliva contains lysozyme **(1)** and the stomach has hydrochloric acid **(1)**; these substances will not reach pathogens inside larger pieces of food **(1)**.

38. The immune system

White blood cells ingest pathogens (phagocytosis) **(1)** and destroy them **(1)**. White blood cells produce antibodies **(1)** that attach to antigens leading to the destruction of the pathogen **(1)**. White blood cells produce antitoxins **(1)**; these attach to poisonous substances and inactivate them **(1)**.

39. Vaccination

1 A vaccine contains a dead or harmless form of a pathogen **(1)**; vaccination means giving a vaccine to cause an immune response in the body **(1)**.

2 Infection can only be transmitted from an infected individual **(1)**; vaccinated people are immune and so cannot become infected **(1)**; if most people are immune, the chances of an unvaccinated person meeting an infected person are very small **(1)**.

40. Antibiotics and painkillers

1 Measles is caused by a virus **(1)**; antibiotics do not kill/affect viruses **(1)**.

2 Antibiotics kill non-resistant strains of bacteria **(1)**; this leaves resistant strains to reproduce/increase in numbers **(1)**; people are more likely be infected by strains which antibiotics cannot treat **(1)**.

41. New medicines

1 digitalis/digoxin from foxgloves **(1)** to treat irregular heartbeat **(1)**

OR

aspirin from willow tree bark **(1)** to treat aches and pains/as a painkiller **(1)**

2 The doctors may unintentionally give away clues to the patients (called observer bias) **(1)**; if the patients are able to work out who receives the drug and who receives the placebo, the trial is less reliable **(1)**.

42. Monoclonal antibodies

Produce monoclonal antibodies that recognise and attach to antigens produced by the cells **(1)**; attach the radioactive substance to the monoclonal antibody molecules **(1)**; inject the antibodies/add the antibodies to a sample of cells **(1)**; the antibodies bind to the cells (delivering the radioactive substance to the cells) **(1)**.

43. Plant disease

The plant material can be examined under a microscope **(1)**; it can be tested to see if it contains harmful bacteria, e.g. by using bacterial culture/monoclonal antibody test kits **(1)**; the soil sample can be analysed to see if it contains enough mineral ions/nitrate ions/magnesium ions for healthy plant growth **(1)**.

44. Plant defences

1 any one from: thorns/hairs/mimicry/leaves which droop or curl when touched **(1)**

2 Cellulose cell walls make it difficult for microorganisms to pass through **(1)**; the waxy leaf cuticle protects against microorganisms **(1)**; layers of dead cells around stems/bark make it difficult for pests/pathogens to penetrate **(1)**.

3 Pests and pathogens damage plants **(1)**; if the plant produces substances that protect it against attack, then the plant will be able to produce more food/grow more rapidly/not be damaged **(1)**.

45. Extended response – Infection and response

*Answer could include the following points to 6 marks:

Benefits to vaccinated children:
- Children become immune to measles, mumps and rubella.
- usually for life
- This protects them against the effects of having these diseases (which can be serious).

Drawbacks to vaccinated children:
- possible side effects
- in small numbers of children.

Benefits to the general population:
- Target may be linked to herd immunity.
- point at which the chance of a non-immunised person coming into a contact with an infected person is very low
- low enough for the pathogen to be lost from the population
- This protects unvaccinated children and adults.

Other points:
- Benefits greatly outweigh the risks.
- None of the years 1988–2014 met the World Health Organization target.
- so benefits vaccinated individuals but not the population as a whole.

46. Photosynthesis

The green parts contain chlorophyll **(1)** which absorbs light/energy for photosynthesis **(1)**; photosynthesis only happens in the green parts **(1)**; starch is made from glucose (in these areas) **(1)**.

47. Limiting factors

Carbon dioxide is the limiting factor because increasing the concentration of carbon dioxide increases the rate of photosynthesis **(1)** and increasing the light intensity makes no difference to the rate **(1)**.

48. Investigating photosynthesis

1 correct values in m **(1)**; correct light intensities **(1)** given to 2 significant figures **(1)**

Distance (cm)	Distance (m)	Light intensity
10	0.10	$1/0.10^2 = 100$
20	0.20	$1/0.20^2 = 25$
30	0.30	$1/0.30^2 = 11$
40	0.40	$1/0.40^2 = 6.3$

2 axes labelled with quantity and unit, and scales chosen so plotted points cover at least 50% **(1)**; all five points plotted $\pm$ ½ square **(1)**; curve of best fit drawn **(1)**; e.g.

As the light intensity increases the rate of bubbling/photosynthesis increases **(1)**; the rate of change becomes less at higher light intensities **(1)**.

49. Respiration

(a) advantage: releases more energy than anaerobic respiration **(1)**; disadvantage: needs oxygen/cannot happen without oxygen/produces carbon dioxide (which must be removed from cells) **(1)**

(b) advantage: does not need oxygen/does not release carbon dioxide **(1)**; disadvantage: releases less energy than aerobic respiration/produces lactic acid **(1)**

50. Responding to exercise

1 Anaerobic respiration occurs/glucose undergoes incomplete oxidation **(1)** and lactic acid builds up in the muscles **(1)**.

2 Anaerobic respiration occurs **(1)**, lactic acid builds up in the muscles **(1)** and oxygen is needed to react with lactic acid (to remove it from cells) **(1)**.

51. Metabolism

1 (the sum of) all the reactions that happen in a cell or in the body **(1)**.

2 amylase/carbohydrase **(1)**; lipase **(1)**; protease **(1)**

52. Extended response – Bioenergetics

*Answer could include the following points to 6 marks:
Increased breathing rate:

- increased muscle contraction
- so more oxygen is needed for respiration
- and more carbon dioxide produced by muscle cells
- rate of gas exchange in the lungs increased
- more oxygen inhaled/diffuses into the blood
- more carbon dioxide exhaled/removed from the body.

Increased heart rate:

- blood is pumped faster around the body/to muscles
- more oxygen delivered to muscles
- more glucose delivered to muscles
- for aerobic respiration
- which releases energy.

Anaerobic respiration:

- reduced need for anaerobic respiration
- so reduces build-up of lactic acid in the muscles.

53. Homeostasis

the regulation of the internal conditions of a cell or organism **(1)** to maintain optimum conditions for function **(1)** in response to internal and external changes **(1)**

54. Neurones

1 Sensory neurones carry electrical impulses from receptor cells to the central nervous system **(1)**; relay neurones link other neurones together and make up the nervous tissue of the central nervous system **(1)**; motor neurones carry nerve impulses from the central nervous system to effectors **(1)**.

2 The axon and dendron are long to transmit impulses over long distances **(1)**; dendrites can receive impulses from receptor cells **(1)**; axon terminals can connect with other neurones **(1)**; myelin sheath insulates the neurone from surrounding neurones/helps the electrical impulse to travel faster **(1)**.

55. Reflex actions

It takes time for the impulses to reach the brain **(1)** and this is longer than it takes for impulses to reach the effector **(1)**.

56. Investigating reaction times

Practice may improve reaction time **(1)**, so practice before starting would affect the results/reaction times **(1)**.

57. The brain

(a) medulla **(1)**

(b) cerebral cortex **(1)**

58. The eye

Circular muscles (of the iris) contract **(1)**; radial muscles (of the iris) relax **(1)**; the pupil constricts/becomes smaller (which lets less light into the eye) **(1)**.

59. Eye defects

1 Near objects are clear with myopia but blurred with hyperopia **(1)**; distant objects are blurred with myopia but clear with hyperopia **(1)**.

2 Hyperopia is caused by the eyeball being too short/the cornea being too flat/ciliary muscles not being strong enough to make the lens thick enough **(1)**, so light focuses behind the retina **(1)**; a convex/converging spectacle lens **(1)** refracts/converges light **(1)**, so the eye can focus it on the retina **(1)**.

60. Thermoregulation

We sweat more when it is hot **(1)**; we need to replace more water lost in sweat **(1)**.

61. Hormones

1 substance secreted by a gland **(1)**; transported in the bloodstream **(1)**; has an effect on a target organ **(1)**

2 Hormonal responses are slower **(1)** and have longer-lasting effects **(1)**.

   or the opposite for nervous responses.

62. Blood glucose regulation

1 (a) pancreas **(1)**

   (b) liver **(1)**

2 A change in concentration causes responses that reverse the change **(1)** so that blood glucose concentration is maintained within a small range **(1)**.

63. Diabetes

More insulin is needed if a big meal is taken rather than a small meal **(1)**; more insulin is needed if a meal contains a lot of sugar/carbohydrates **(1)**; less insulin is needed if the person is exercising (because glucose is used up in respiration) **(1)**.

64. Controlling water balance

Protein molecules are too large **(1)** to pass through the pores/holes in the filter/membrane **(1)**.

65. ADH and the kidneys

As alcohol inhibits ADH production, less ADH will be secreted **(1)**, the tubules will become less permeable **(1)** so less water will be reabsorbed **(1)**, and a large volume of dilute urine will be produced **(1)**.

66. Kidney treatments

The concentration of urea/salts in the dialysis fluid will increase in the machine **(1)**, reducing the concentration difference/gradient/rate of diffusion **(1)**, so fresh dialysis fluid maintains a high concentration difference/gradient **(1)**.

67. Reproductive hormones

If an egg is fertilised, the uterus lining is maintained **(1)** and menstruation does not happen **(1)**.

68. Control of menstrual cycle

The levels of FSH and LH will fall/remain low **(1)**; progesterone inhibits their release **(1)**; the lining of the uterus remains thick/menstruation does not occur **(1)**; progesterone maintains the lining **(1)**.

69. Contraception

two from the following:

- Use oral contraceptive (1) which contains hormones that inhibit FSH production/stop eggs maturing (1).
- Use injection/implant/skin patch containing progesterone (1) which prevents egg maturation/release (1).
- Use a diaphragm (1) which fits over the cervix/stops sperm entering the uterus (1).
- Use an intrauterine device (IUD) (1) which stops sperm and eggs surviving in the uterus (1).
- Use natural family planning (1) which involves abstaining from intercourse when an egg may be in the oviduct (1).

70. Treating infertility

1 three from the following for 1 mark each: emotionally stressful; physically stressful; can lead to multiple births; low success rates

2 to make sure that too many eggs do not mature/are not released at once (1), because this could lead to multiple births (1) which are a risk to babies/mother (1)

71. Adrenaline and thyroxine

An overactive thyroid gland produces more thyroxine than normal (1); the rate of respiration is greater than normal (1), using up glucose/food faster than normal (1); the thyroid gland is located in the neck (1).

72. Plant hormones

1 Auxins stimulate cell elongation in shoot cells (1); auxins inhibit cell elongation in root cells (1).

2 Negative phototropism makes roots grow towards the dark (1) and positive gravitropism/geotropism makes roots grow downwards (1), both of which are likely to be towards the soil (1); roots anchor plants in soil/absorb water and minerals from the soil (1).

73. Investigating plant responses

1 to increase the validity of the experiment/results (1); to allow a mean to be calculated (1)

2 The first few millimetres of the seedlings will not be measured (1), so the heights will all be too small (1) by the same length/zero error (1).

74. Uses of plant hormones

1 two from the following for 1 mark each: (selective) weedkillers; rooting powder; promoting growth in tissue culture

2 Bananas release ethene (1); ethene promotes ripening (1).

75. Extended response – Homeostasis and response

*Answer could include the following points to 6 marks:

- Receptor cells in the eye detect a stimulus.
- The stimulus is the bright light/increase in light intensity.
- Receptor cells pass impulses to sensory neurone.
- Sensory neurone passes impulses to relay neurone.
- Relay neurone is in the spinal cord/central nervous system/CNS.
- Relay neurone passes impulses to motor neurone.
- Motor neurone passes impulses to the effector/muscle in the iris.
- Muscle contracts/makes pupil smaller.
- synapses between neurones
- Chemicals carry impulses across the synapse.

76. Meiosis

Haploid cells have half the number of chromosomes/one set of chromosomes, but diploid cells have the normal number of chromosomes/two sets of chromosomes (1); haploid cells are formed by meiosis but diploid cells are formed by fertilisation (1) and by mitosis of body cells (1); gametes are haploid cells but body cells are diploid cells (1).

77. Sexual and asexual reproduction

1 Strawberry plants produce runners with small plants along them (1), and daffodils produce underground bulbs which divide to form new bulbs (1), by mitosis/asexual reproduction (1).

2 Non-identical gametes (1) are produced by meiosis (1); these join/fuse to form a new individual (1), which involves mixing of the genetic material (1).

78. DNA and the genome

DNA is a polymer made up of repeating nucleotide units (1), made of two strands of phosphate and sugar groups (1) in a double helix (1); a base is joined to each sugar (1); the four bases are A, C, G and T (1).

79. Protein synthesis

DNA is used to produce a template (1) which binds to a ribosome (1); a carrier molecule brings specific amino acids (1); each amino acid is added to the growing protein chain (1) in the correct order (1); the completed protein folds into a unique shape (1).

80. Genetic terms

The R allele is dominant (1) because it is expressed if there are two copies of the allele or only one (1); the r allele is recessive (1) because it is only expressed if there are two copies of the allele (1).

81. Genetic crosses

(a) Punnett square diagram drawn, e.g.

Gametes	B	B
b	Bb	Bb
b	Bb	Bb

parents' alleles correct (1); offspring alleles correct (1)

(b) There is no chance (0/4, 0, 0%) (1) because all the offspring will be heterozygous (1) and b/brown is only expressed if there are two copies of the allele (1).

82. Family trees

(a) 10 is tt (1) because they are not a taster (and PTC tasting is dominant) (1); 9 is Tt (1) because some of his children are not tasters (1).

(b) Punnett square diagram drawn, e.g.

Gametes	T	t
t	Tt	tt
t	Tt	tt

parents' alleles correct (1); offspring alleles correct (1)

The tasters are Tt (1), so the probability is 1/2 (or 0.5/50%/1 in 2/1:1) (1).

83. Inheritance

1 Females have the same sex chromosomes/XX chromosomes (1), so egg cells (formed by meiosis) must contain an X chromosome (1).

2 two from the following for 1 mark each:
- Polydactyly is due to a dominant allele, but cystic fibrosis is due to a recessive allele.

- People can have polydactyly if they inherit one or two of these alleles, but people must inherit two of these alleles to have cystic fibrosis.
- People can be carriers of cystic fibrosis/have the allele but do not have the disorder, but there are no carriers of polydactyly.

84. Variation and evolution

1 Identical twins have the same alleles/genotype because they come from the same fertilised egg (1), so characteristics controlled by genes will be the same (1); characteristics affected by the environment may be different between the twins (1).

2 Mutation occurs that allows resistance (1); lice with the mutant allele are not killed by the treatments (1); they reproduce and pass on the allele to their offspring (1); the proportion of lice with the mutant allele increases over time (1).

85. Selective breeding

Select the sheep in the flock that have the most wool (1); breed from these sheep (1); choose sheep with the most wool from the offspring and breed from them (1); repeat this process over many generations (1).

86. Genetic engineering

1 An organism's genome is modified (1) to give a desired characteristic (1) by introducing a gene from another organism (1).

2 Cut the gene for herbicide resistance from (the genome of) an organism (1); transfer the gene into crop plant cells (1); grow the cells into plants (1).

87. Stages in genetic engineering

an organism/part of an organism that can enter a cell (1) and insert DNA into the cell's DNA (1)

88. Cloning

Embryo transplant, because several embryo cells can be separated for transplant (1), but in adult cell cloning each body cell only produces one embryo (1).

89. Darwin and Lamarck

A giraffe with a longer neck can reach food high up (1); this giraffe is more likely to get enough food to survive to reproduce (1); its offspring inherit the long neck (1).

90. Speciation

Birds from species in Australia reached New Zealand and so became separated from their original population (1); natural selection favoured different variations in characteristics (1), including flightlessness (as there are no natural predators in New Zealand) (1); these new species are so different from the original species that they cannot interbreed to produce fertile offspring (1).

91. Mendel

two from the following:
- started with pure-breeding parent plants (1) which made the results more repeatable (1)
- made the crosses by hand (1) so he could clearly link the parent cross to the results (1)
- repeated each test many times (1) to average out chance/random variation in the results (1).

92. Fossils

1 Soft-bodied animals do not form fossils as frequently/easily as the bones of vertebrates. (1)

2 These animals are adapted to living in very cold conditions (1); if the climate gets warmer quickly, they may not be able to adapt fast enough to the new conditions, and so will die out (1).

93. Resistant bacteria

(a) to kill bacteria on the hands (1), so these cannot be spread from place to place in the hospital (1)

(b) to see if the patient is carrying MRSA (1); they can be isolated/treated if they do (1)

94. Classification

The chameleon shares more characteristics with the viper, because they are in the same class (1), which is a smaller group of organisms than a phylum (1).

95. Evolutionary trees

(a) Cows and pigs are more similar to each other (1) than they are to humans (1).

(b) cow and pig (1), because the common ancestor for cow and hedgehog lived further back in time than the common ancestor for cow and pig (1)

96. Extended response – Inheritance, variation and evolution

*Answer could include the following points to 6 marks:
Similarities:
- Both take many generations.
- Both involve selection of certain variants.
- Both involve starting with mixed populations.
- Both involve reproduction using the individuals with a certain characteristic.

Differences:
- Natural selection does not need human intervention but selective breeding does.
- Selective breeding involves aiming for a desired characteristic but natural selection does not.
- Natural selection works over many more generations than selective breeding.

97. Ecosystems

1 organism: a single living individual (1); population: all the organisms of the same species in a habitat (1); community: the populations of all the different species in a habitat (1); ecosystem: the interaction of a community of living organisms with the non-living parts of their environment (1)

2 There is very little light under the trees (1); there is not enough light for other plants to photosynthesise (1).

98. Interdependence

two possible results, each with a reason, e.g.
- The population of sparrows would increase (1) because sparrowhawks would no longer eat them (1).
- The population of foxes would increase (1) because there would be more fieldmice (1).
- The population of rabbits would increase (1) because foxes would eat more fieldmice instead (1).
- The population of grass would decrease (1) because there would be more sparrows (1).

99. Adaptation

1 The fur is camouflage (against predators) (1), so it needs to be white against the snow in the winter and brown against the ground/plants in the summer (1).

2 Deep in the ocean it is dark (1), so the light patterns can attract a mate/prey or scare away predators (1).

100. Food chains

Zebra numbers rise because there is plenty of food for them (e.g. grass/leaves) and few lions, so more zebras survive to reproduce successfully (1); more zebras means more food for the lions, so more lions survive to reproduce successfully (1); as the number of lions rises, they eat a greater proportion of zebras, so the number of zebras falls (1); there is less food for the lions, so the number of lions falls (1).

101. Fieldwork techniques

mean number of dandelions in a quadrat = 142/10 = 14.2 (1)

area of field = 100 × 200 = 20 000 m^2 (1)

estimated population = 14.2 × 20 000/1 = 284 000 (1)

102. Field investigations

(a) The number of dandelions increases as the distance from the tree increases. (1)

(b) The light intensity also increases as the distance from the tree increases (1); plants need light for photosynthesis (1); if there is more light available, the dandelion plants can make more glucose/can grow better (1).

103. Cycling materials

(a) respiration (1); combustion (1)

(b) photosynthesis (1)

104. Decomposition

1 Warmth from the sun (1) increases the rate of growth of microorganisms so they break down materials faster (1).

2 Air/oxygen is added to the compost when it is turned over (1), increasing the rate of (aerobic) respiration by decay microorganisms (1).

105. Investigating decay

(a) The time taken decreases if the rate of decay increases (1); the rate of reaction increases as the temperature increases (1).

(b) 55 °C is above the optimum temperature of the lipase (1); at high temperatures the lipase/enzyme will be denatured (1), so decay does not happen/the lipase does not break down the milk and the milk does not become acidic (1).

106. Environmental change

(a) temperature (1); availability of water (1); composition of atmospheric gases (1)

(b) seasonal changes (1); geographical changes (1); human interaction/activities (1)

107. Waste management

Human population is increasing (1); more food is needed (1) so more pesticides and herbicides are needed to kill pests and weeds (1).

108. Deforestation

carbon dioxide released by burning/combustion of trees (1); more carbon dioxide released by increased activity/respiration of decay microorganisms (1); less carbon dioxide removed from the air by photosynthesis/carbon dioxide removed more slowly/not 'locked up' in wood for as long (1)

109. Global warming

The organisms that cause/transmit tropical diseases can only live where it is warm (1); if the climate gets warmer in other places, these organisms may be able to survive there (1) so people in those places may catch the diseases (1).

110. Maintaining biodiversity

Hedges provide food/shelter/nesting places for insects/birds (1); insects/birds provide food for other animals (1) so the number of different species increases (1).

111. Trophic levels

Not all ingested material is absorbed (1); some is egested as faeces (1); some material lost as waste (e.g. carbon dioxide/water in respiration; water/urea in urine) (1); glucose lost in respiration (1).

112. Food security

Exports are not likely to affect food security because many people in India are vegetarian (1); the buffalo are raised to supply the dairy industry (1), so selling their meat makes use of a resource that would not be widely used locally (1).

113. Farming techniques

Usually the lowest level of a food chain is the most efficient for gathering food for humans (1) as biomass is lost moving from one trophic level to the next (1), but it might be difficult to collect large amounts of plankton to make into food (1); catching and eating sardines will be more efficient than catching and eating tuna (1).

114. Sustainable fisheries

(a) The dead fish were fish that were too small to sell or be caught (legally). (1)

(b) The mesh size was a minimum size (1) which let small fish escape/not be caught (1).

115. Biotechnology and food

growing *Fusarium* in fermenters (1) to produce large amounts of mycoprotein (1); using genetic engineering/modification (1) to produce crops with high yields/nutritional value (1)

116. Extended response – Ecology

*Answer could include the following points to 6 marks:

Features of factory farming (could be mentioned in advantages or in disadvantages):

- Pigs are kept in small pens.
- The temperature is controlled.
- High-protein foods are used.
- Antibiotics may be given.

Advantages:

- Less energy is transferred from the pigs to the surroundings.
- So pigs need less food.
- High-protein foods encourage faster growth.
- More pigs can be kept in the same area.
- Predators cannot harm/kill the pigs.
- Antibiotics control disease/increase growth.

Disadvantages:

- The pigs' movement is restricted.
- ethical concerns, e.g. pigs are stressed/overcrowded; people may think it is cruel
- High-protein foods might have been processed to feed people instead.
- Agricultural use of antibiotics may produce resistant strains of bacteria.
- Heating uses energy/fuel.

Conclusion:

- for/against factory farming, with reason(s) given.

Your own notes

Published by Pearson Education Limited, 80 Strand, London, WC2R 0RL.

www.pearsonschoolsandfecolleges.co.uk

Text and illustrations © Pearson Education Ltd 2017
Typeset, illustrated and produced by Phoenix Photosetting
Cover illustration by Miriam Sturdee

The right of Nigel Saunders to be identified as author of this work has been asserted by him in accordance with the Copyright, Designs and Patents Act 1988.

Content written by Sue Kearsey and Pauline Lowrie is included.

First published 2017

20 19 18 17
10 9 8 7 6 5 4 3 2 1

British Library Cataloguing in Publication Data
A catalogue record for this book is available from the British Library

ISBN 978 1 292 13503 8

Printed in Slovakia by Neografia.

Acknowledgements
The publisher would like to thank the following for their kind permission to reproduce their photographs:
(Key: b-bottom; c-centre; l-left; r-right; t-top)
123RF.com: kukhunthod 21; **Science Photo Library Ltd:** Herve Conge, ISM 6; **Shutterstock.com:** Buquet Christophe 43, Christopher Boswell 108, mark higgins 92, xpixel 57
All other images © Pearson Education

Note from the publisher
Pearson has robust editorial processes, including answer and fact checks, to ensure the accuracy of the content in this publication, and every effort is made to ensure this publication is free of errors. We are, however, only human, and occasionally errors do occur. Pearson is not liable for any misunderstandings that arise as a result of errors in this publication, but it is our priority to ensure that the content is accurate. If you spot an error, please do contact us at resourcescorrections@pearson.com so we can make sure it is corrected.

$$1 = \frac{1}{30}\left(\sin\left(-90 - b\right)\right) + -1$$

A waxy